Mathematical Problem Solving

Workbook 4

Strategy for Solving Real-World Problems

Satya Pradhan

ISBN: 1541377443

ISBN 13: 978-1541377448

Contents

Introduction

Having strong problem-solving skills can make a huge difference in one's career in the modern knowledge-based economy. Problems are at the center of what one does at work every day. Whether one is developing a vaccine for the winter flu, creating an antivirus for the Internet, delivering lifesaving drugs to remote villages, maximizing profits for a company, or understanding the complex structure of our universe, problems are an integral part of our everyday lives. So being an effective and confident problem solver is really important to one's success. Much of that confidence comes from having a good understanding of strategy and the tools to use when approaching a problem. Therefore, it is essential for students to develop the skills and techniques for problem solving from an early age, when they are in elementary school.

Conceptual understanding, procedural and computational skills, and application of concepts to real-life problems are the three pillars of a mathematics education. Conceptual understanding involves knowing what to do; procedural fluency requires knowing how to do it; and problem solving focuses on solving a wide variety of complex real-life problems using mathematical knowledge. Mathematical skills have been taught in school in this order: maximum emphasis is placed on the understanding of math concepts and computational skills followed by the application of the concepts to real-world problems. However, the real-life problem solving requires students to apply these concepts in the exact opposite order, starting with understanding the problem, then finding the mathematical concepts required to solve the problem, and finally choosing the method that best solves the problem.

Mathematical problem solving is often taught as a way to reinforce mathematical concepts, which misses the importance of strategic thinking while solving a problem. Many research articles and books have been written emphasizing the importance of problem-solving strategy. However, the burden of teaching problem-solving strategy is left mostly to teachers and parents, who are expected to develop their own curriculums and lesson plans for the complex topic of strategy and then teach it to students.

This book presents several problem-solving strategies that can easily be used by teachers and parents to teach the subject. The first two chapters present the concepts of number operations and the basic problem-solving strategies listed below:
- solving one-step problems
- solving multistep problems
- solving problems working backward
- formulating problems with variables and equations
- solving problems using variables

Then the concept of the unitary method is presented in chapter 3. The remaining chapters present lessons on different problem types. The objective is to teach students how to start with a problem statement, understand the problem, and then solve it with a known mathematical procedure.

There are many different problem types that students will encounter in their careers. We have selected the following problem types that are appropriate for students in fourth grade:
- number problems
- age problems
- time and distance problems
- money problems
- work problems
- mixture problems

Each lesson in the workbook is classified as (*), (**), or (***), depending on the level of difficulty, and each starts with few examples showing how to solve a particular type of problem. This is followed by a number of problems of this type. Students are expected to know basic computational skills in order to solve the problems in this workbook.

Notes to Parents, Teachers, and Tutors

As a parent, you can use this workbook to teach problem-solving techniques to your child without any teaching experience. The first three chapters present the basic concepts and should be taught first. If students are already familiar with these concepts, these chapters can be skipped. All other chapters are independent of one another and can be taught in any order.

As a schoolteacher, you can easily integrate this workbook into the school curriculum by choosing appropriate lessons to teach along with the curriculum.

Private tutors and after-school learning centers can use this workbook to offer special classes on mathematical problem solving or as part of other math-enrichment programs. It is suggested to teach two or three lessons a week, using the example questions given at the beginning of each lesson and give the other questions in the lessons as homework.

Conceptual understanding of mathematical problem solving is the main focus of this book. Therefore, we encourage students to use calculators to solve the numerical expressions. This will allow them to take less time for numerical calculation so they can instead focus on understanding concepts.

Answer Keys

Answer keys for all questions in this book are available online. You can download the PDF file for the answer keys at www.ilecy.com/BookAnswers

Feedback

We are always looking for feedback from students, parents, and teachers to make this book better. Please send your comments, testimonials, or suggestions for improvement to mathPS100@gmail.com.

Acknowledgments

My sincere appreciation and thanks to the following people for their feedback and suggestions while teaching these lessons as part of an after-school math enrichment program: Susie Bierman, Reynaldo Lorenzana, Shalini Kinger, Arun Sahoo, and Alicia Lopez. I would also like to acknowledge the help of Soumya Sahoo, Kallala Giri, and Kamalesh Parhi during the preparation of this workbook. My sincere appreciation to my wife, Nishi, for her support while I was working on this book along with my hectic full-time job in Silicon Valley, California. Special thanks to my son, Sougat, and daughter, Sarika, for their valuable feedback on different lessons. They were my first reviewers, connecting the lessons to their classrooms at school.

Assessment

Note: Some of the assessment questions may be challenging at the beginning of fourth grade.

Chapter 1:

1. What is the operation keyword(s) in the following sentence?

   The sum of 15 and 35 is 50.

   Answer: _____

2. What is the math sentence for the following expression?

   The double of 25 added to 75

   (a) $75 - (2 + 25)$
   (b) $(25 \div 2) + 75$
   (c) $(2 \times 25) + 75$
   (d) $25 - (2 \times 75)$

   Answer: _____

Chapter 2:

3. Mr. Hasley bought some fruit. He paid $7.50 for apples and $9.50 for strawberries. He was left with the same amount of money as he spent at the fruit shop. How much did he have at the beginning?

   Answer: _____

4. Charles had 40 candies. He divided the candies equally among 6 kids and kept the remaining candies with him. How many candies did Charles keep with him?

   Answer: ____ _____
   unit

5. Review the question given below, and choose the best choice for the information available.

   Paul bought 15 cookies and 3 pastries. He ate 6 cookies. How many total cookies and pastries does he have left?

   (a) Too much information
   (b) The right amount of information
   (c) Too little information

   Answer: _____

Chapter 3:

6. 12 workers can construct a road in 5 days. How many days will it take for 15 workers to construct the road?

   Answer: ____ _____
   unit

7. Mary and 2 of her friends take 1 hour to clean the school garden. How long will it take Mary to clean the garden by herself?

   Answer: ____ _____
   unit

8. 5 students eat 20 oranges in a week. How many students will eat 48 oranges in a week?

   Answer: ____ _____
   unit

Chapter 4:

9. What is the sum of the place values of 8, 2, and 5 in 8,251?

    Answer: _____

10. I am a number 60 more than the smallest possible three-digit number that can be made using the digits 8, 2, and 5. What number am I?

    Answer: _____

11. I am the smallest number that never repeats a digit and is greater than 40,000. What number am I?

    Answer: _____

12. Kim wrote 48 stories and 60 poems in a month. About how many hundreds total did he write?

    Answer: _____

13. If you replace the thousands digit by the sum of the tens and the hundreds digit in the number 4,625, what will be the new number?

    Answer: _____

14. Write 9,764 in expanded form, and find the missing number in the following math sentence.

    9,764 = 9,000 + 700 + _____ + 4

    Answer: _____

Chapter 5:

15. Kelly was 7 years old in 2012. How old will she be in 2021?

    Answer: ____ _____
    unit

16. The difference between Donald's and Brian's ages is 11 years. What will be the difference in their ages 5 years from now?

    Answer: ____ _____
    unit

17. Jay's current age is one-third of his uncle's age. If his uncle's age is 45, what is Jay's current age?

    Answer: ____ _____
    unit

18. Carl's current age is three-fourths of his brother's age. His brother's age is 16. What is Carl's age?

    Answer: ____ _____
    unit

19. Hazel is currently 18 years old. Her mother is 3 times as old as Hazel. What is the sum of their ages?

    Answer: ____ _____
    unit

20. Rob's age is two-thirds of his sister's age. His sister's age is 15. Find the difference between their ages.

    Answer: ____ _____
    unit

Chapter 6:

21. If it is 5:15 p.m. now, what will be the time in 4 hours?

    Answer: _____ _____
    unit

22. A movie started at 9:35 p.m. and ended at 11:35 p.m. How long was the movie in hours?

    Answer: _____ _____
    unit

23. If July 18 is on Monday, what day of the week will it be on July 22?

    Answer: _____

24. Martin is visiting his friend in a different city and wants to stay for 8 days at his friend's house. If he starts on March 12, what day will he return?

    Answer: _____ _____
    unit

25. Adam's family had to drive 100 kilometers to go on a picnic. They started from their home at 7:30 a.m. and reached the place of picnic at 9:30 a.m. What was their speed in kilometers per hour?

    Answer: _____ _____
    unit

26. Bradley runs 160 meters in 1 minute. If he has to run for 1 kilometer and 280 meters, how much time will he take?

    Answer: _____ _____
    unit

Chapter 7:

27. Tanya ordered 3 hot coffees and 6 glasses of juice and paid a total of $15.00. If the cost of each hot coffee is $2.00, what is the cost of each glass of juice?

    Answer: _____

28. Ricky wants to buy a guitar that costs $50.00. He also needs to buy headphones that cost $12.50 and a wallet that costs $9.75. How much money does he need in total?

    Answer: _____

29. Lora deposited $560.00 in the bank. After two years she received a balance of $625.00. How much interest did she get in two years?

    Answer: _____

30. Lewis bought 24 novels for his store and paid a total of $96.00. How much did he pay for each novel?

    Answer: _____

31. Jenny bought a watch and sold it for $45.00. If she experienced a loss of $13.00, how much did she pay to buy the watch?

    Answer: _____

Chapter 8:

32. 8 engineers can complete a survey in 3 hours. How many hours will 1 engineer need for the same survey?

Answer: _____ _____
unit

33. 5 boys can arrange 120 chairs in 2 hours. How many chairs can they arrange in 3 hours?

Answer: _____ _____
unit

34. 4 people can assemble a water tank in 6 hours. If 2 more people join, how long will they take to complete the assembly?

Answer: _____ _____
unit

35. A group of 8 ants can collect 160 grains in one day. How many ants can collect 240 grains in the same time?

Answer: _____ _____
unit

36. Max takes 52 minutes to complete an assignment. How long will he take to complete three-fourths of the assignment?

Answer: _____ _____
unit

Chapter-9:

37. Basket 1 has 14 mangoes and 15 bananas. Basket 2 has 20 mangoes and 19 bananas. If we mix all the mangoes from both the baskets together, what fraction of the total fruit is mangoes?

Answer: _____

38. Two jars (Jar A and Jar B) have 36 liters of solution in each of them. Jar A is one-third water and Jar B is one-fourth water. What is the total quantity of water in both the jars?

Answer: _____ _____
unit

39. A pipe can empty a drum in 20 minutes. If we start emptying a full drum, what fraction of the drum will still be full after 15 minutes?

Answer: _____

40. Tap A can fill 1,000 liters of water into a container in 1 hour, and Tap B can empty 750 liters of water from the container in 1 hour. If both the taps are opened together, what quantity of water will be filled in 1 hour?

Answer: _____ _____
unit

1. Mathematical Operations

1.1 Addition and Subtraction Keywords (*)

Example 1:

What is the addition or subtraction keyword(s) in the following expression?

Sum of 13 and 18

Solution:

In this expression, *sum of* are the addition keywords.

Example 2:

What operation will you use for the keyword *remainder*?
(a) Subtraction
(b) Addition
(c) None of the above

Solution:

The keyword *remainder* is a *subtraction* keyword. So the answer is (a).

Write or choose the letter of the answer.

1. What is the operation keyword(s) in the following sentence?

    How many total mangoes and apples are there in a bag?

    Answer: _____

2. What operation will you use for the keywords *total of*?
    (a) Addition
    (b) Subtraction
    (c) Both (a) and (b)
    (d) None of the above

    Answer: _____

3. What is the operation keyword(s) in the following sentence?

    Mr. White had a loss of $45.00 in his business.

    Answer: _____

4. What operation will you use for the keyword(s) in question 3?
    (a) Addition
    (b) Subtraction
    (c) Both (a) and (b)
    (d) None of the above

    Answer: _____

5. What operation will you use for the keyword *less*?
    (a) Addition
    (b) Subtraction
    (c) Both (a) and (b)
    (d) None of the above

    Answer: _____

6. What is the operation keyword(s) in the following expression?

    35 less than 43

    Answer: _____

Write or choose the letter of the answer.

7. What is the operation keyword(s) in the following sentence?

    Lora gained 7 pounds last month.

    Answer: _____

8. What operation will you use for the keyword(s) in question 7?
    (a) Addition
    (b) Subtraction
    (c) Both (a) and (b)
    (d) None of the above

    Answer: _____

9. What is the operation keyword(s) in the following sentence?

    Bob has 12 more candies than Bill has.

    Answer: _____

10. What operation will you use for the keyword *fewer*?
    (a) Addition
    (b) Subtraction
    (c) Both (a) and (b)
    (d) None of the above

    Answer: _____

11. What is the operation keyword(s) in the following expression?

    Sum of 5 and 12

    Answer: _____

12. What operation will you use for the keyword *minus*?
    (a) Addition
    (b) Subtraction
    (c) Both (a) and (b)
    (d) None of the above

    Answer: _____

13. What operation will you use for the keywords *increase of*?
    (a) Addition
    (b) Subtraction
    (c) Both (a) and (b)
    (d) None of the above

    Answer: _____

14. What operation will you use for the keyword *plus*?
    (a) Addition
    (b) Subtraction
    (c) Both (a) and (b)
    (d) None of the above

    Answer: _____

15. What operation will you use for the keywords *take away*?
    (a) Addition
    (b) Subtraction
    (c) Both (a) and (b)
    (d) None of the above

    Answer: _____

1.2 Multiplication and Division Keywords (*)

Example 1:

What operation will you use for the keyword *every*?

(a) Multiplication
(b) Division
(c) None of the above

Solution:

Every is a *division* keyword. So the answer is (b).

Example 2:

What is the multiplication or division keyword(s) in the following sentence?

Multiply 12 and 8.

Solution:

In this sentence, *multiply* is the multiplication keyword.

Write or choose the letter of the answer.

1. What is the operation keyword(s) in the following sentence?

Find the double of 9.

Answer: _____

2. What operation will you use for the keyword(s) in question 1?

(a) Multiplication
(b) Division
(c) None of the above

Answer: _____

3. What operation will you use for the keyword *per*?

(a) Multiplication
(b) Division
(c) None of the above

Answer: _____

4. What operation will you use for the keywords *product of*?

(a) Division
(b) Multiplication
(c) None of the above

Answer: _____

5. What operation will you use for the keywords *divide by*?

(a) Multiplication
(b) Division
(c) None of the above

Answer: _____

6. What operation will you use for the keyword *twice*?

(a) Multiplication
(b) Division
(c) None of the above

Answer: _____

Write or choose the letter of the answer.

7. What operation will you use for the keywords *quotient of*?
    (a) Multiplication
    (b) Division
    (c) None of the above

    Answer: _____

8. What is the operation keyword(s) in the following sentence?

    Divide 21 by 7.

    Answer: _____

9. What operation will you use for the keywords *7 days*?
    (a) Multiplication
    (b) Division
    (c) None of the above

    Answer: _____

10. What operation will you use for the keywords *fraction of*?
    (a) Multiplication
    (b) Division
    (c) None of the above

    Answer: _____

11. What is the operation keyword(s) in the following expression?

    Product of 5 and 2

    Answer: _____

12. What operation will you use for the keyword(s) in question 11?
    (a) Division
    (b) Multiplication
    (c) None of the above

    Answer: _____

13. What is the operation keyword(s) in the following sentence?

    Andy spent 20% of the salary for his sister's admission.

    Answer: _____

14. What is the operation keyword(s) in the following expression?

    25 divided by 5

    Answer: _____

15. What operation will you use for the keywords *distributed equally*?
    (a) Multiplication
    (b) Division
    (c) None of the above

    Answer: _____

16. What operation will you use for the keyword *times*?
    (a) Multiplication
    (b) Division
    (c) None of the above

    Answer: _____

1.3 Operation Keywords 1 (*)

Example 1:

What is the operation keyword(s) in the following sentence?

John had an increase of $230.00 in his salary.

Solution:

In this sentence, *increase of* are the *addition* keywords.

Example 2:

What operation will you use for the keywords *divided equally*?

(a) Division
(b) Addition
(c) Subtraction
(d) Multiplication

Solution:

The keywords *divided equally* are *division* keywords. So the answer is (a).

Write or choose the letter of the answer.

1. What is the operation keyword(s) in the following sentence?

Manoj has a loss of $30.00.

Answer: _____

2. What operation will you use for the keyword *raise*?
   (a) Subtraction
   (b) Multiplication
   (c) Addition
   (d) Division

Answer: _____

3. What is the operation keyword(s) in the following sentence?

Find the product of 2.3 and 7.

Answer: _____

4. What operation will you use for the keywords *take away*?
   (a) Division
   (b) Addition
   (c) Subtraction
   (d) Multiplication

Answer: _____

5. What is the operation keyword(s) in the following sentence?

Bob sold a motorbike for $600.00 and gained 8%.

Answer: _____

6. What operation will you use for the keyword *triple*?
   (a) Addition
   (b) Division
   (c) Multiplication
   (d) Subtraction

Answer: _____

Write or choose the letter of the answer.

7. What is the operation keyword(s) in the following sentence?

    Lisa made a profit of one-third of $400.00.

    Answer: _____

8. What operation will you use for the keyword *more*?
   (a) Division
   (b) Multiplication
   (c) Subtraction
   (d) Addition

    Answer: _____

9. What operation will you use for the keywords *distributed equally*?
   (a) Addition
   (b) Division
   (c) Subtraction
   (d) Multiplication

    Answer: _____

10. What is the operation keyword(s) in the following expression?

    100 sweets per 10 plates

    Answer: _____

11. What is the operation keyword(s) in the following sentence?

    Sam saves 30% of his salary every month.

    Answer: _____

12. What is the operation keyword(s) in the following sentence?

    Find the difference between 9 and 4.

    Answer: _____

13. What operation will you use for the keyword(s) in question 12?
    (a) Division
    (b) Subtraction
    (c) Addition
    (d) Multiplication

    Answer: _____

14. What operation will you use for the keyword *times*?
    (a) Multiplication
    (b) Subtraction
    (c) Addition
    (d) Division

    Answer: _____

15. What is the operation keyword(s) in the following expression?

    20 divided by 5

    Answer: _____

16. What operation will you use for the keyword(s) in question 15?
    (a) Division
    (b) Multiplication
    (c) Addition
    (d) Subtraction

    Answer: _____

1.4 Operation Keywords 2 (*)

Example 1:

What is the operation keyword(s) in the following sentence?

Liza is 3 times as old as Neha.

Solution:

In this sentence, *times* is the *multiplication* keyword.

Example 2:

What operation will you use for the keyword *less*?
(a) Addition
(b) Multiplication
(c) Subtraction
(d) Division

Solution:

The keyword *less* is a *subtraction* keyword. So the answer is (c).

Write or choose the letter of the answer.

1. What is the operation keyword(s) in the following sentence?

50 ribbons are divided equally in 5 boxes.

Answer: _____

2. What operation will you use for the keyword(s) in question 1?
(a) Multiplication
(b) Division
(c) Addition
(d) Subtraction

Answer: _____

3. What is the operation keyword(s) in the following sentence?

Twice of 2 is 4.

Answer: _____

4. What operation will you use for the keyword(s) in question 3?
(a) Division
(b) Subtraction
(c) Addition
(d) Multiplication

Answer: _____

5. What operation will you use for the keywords *product of*?
(a) Subtraction
(b) Multiplication
(c) Addition
(d) Division

Answer: _____

6. What operation will you use for the keywords *sum of*?
(a) Addition
(b) Multiplication
(c) Subtraction
(d) Division

Answer: _____

Write or choose the letter of the answer.

7. What operation will you use for the keywords *quotient of*?
   (a) Addition
   (b) Division
   (c) Subtraction
   (d) Multiplication

   Answer: _____

8. What is the operation keyword(s) in the following expression?

   35% of 180

   Answer: _____

9. What operation will you use for the keyword(s) in question 8?
   (a) Addition
   (b) Multiplication
   (c) Division
   (d) Subtraction

   Answer: _____

10. What operation will you use for the keywords *total of*?
    (a) Subtraction
    (b) Multiplication
    (c) Division
    (d) Addition

    Answer: _____

11. What is the operation keyword(s) in the following expression?

    Two-fifths of 270

    Answer: _____

12. What is the operation keyword(s) in the following sentences?

    Lincon had 3 pens. He bought 4 more pens.

    Answer: _____

13. What operation will you use for the keyword(s) in question 12?
    (a) Multiplication
    (b) Subtraction
    (c) Division
    (d) Addition

    Answer: _____

14. What operation will you use for the keyword *double*?
    (a) Subtraction
    (b) Division
    (c) Addition
    (d) Multiplication

    Answer: _____

15. What operation will you use for the keyword *every*?
    (a) Addition
    (b) Multiplication
    (c) Subtraction
    (d) Division

    Answer: _____

16. What is the operation keyword(s) in the following sentence?

    Tanya spent 60% of her salary.

    Answer: _____

1.5 Write a Math Expression (*)

Example 1:

What is the math sentence for the following problem?

1 book costs $5.00. What is the cost of 4 books?

(a) 5×4
(b) $4 \div 5$
(c) $4 + 5$
(d) None of the above

Solution:

The following information is given:

Cost of 1 book = $5.00
Number of books = 4

We can answer the question using the following equation:

(cost of 4 books) = (cost of 1 book) × (numbers of books)

$= 5 \times 4$

So the answer is (a).

Example 2:

What is the operation keyword(s) in the following expression?

12.75 taken away from 22

Solution:

In this expression, *taken away* are the subtraction keywords.

Write or choose the letter of the answer.

1. What is the math sentence for the following problem?

1 coffee mug costs $3.00. What is the cost of 8 coffee mugs?

(a) $3 \div 8$
(b) 8×3
(c) $8 + 3$
(d) None of the above

Answer: _____

2. What is the operation keyword(s) in the following sentence?

A company gained 30% of its annual profit.

Answer: _____

3. What is the operation keyword(s) in the following expression?

Sum of 26 and 42

Answer: _____

Write or choose the letter of the answer.

4. 1 pizza costs $7.00. Which math sentence will you use to find the cost of 9 pizzas?
   (a) 7 + 9
   (b) 7 × 9
   (c) 9 ÷ 7
   (d) All of the above

   Answer: _____

5. What is the operation keyword(s) in the following sentence?

   Nancy makes 5 toys more than Joy.
   (a) than
   (b) Nancy
   (c) more
   (d) None of the above

   Answer: _____

6. What is the operation keyword(s) in the following sentence?

   Loni spent $12.00 from the product of $10.00 and $8.00.

   Answer: _____

7. Which math sentence will you use for the following sentence?

   Raman ate 8 candies from the total of 15 candies.
   (a) 8 + 15
   (b) 15 − 8
   (c) 15 ÷ 8
   (d) All of the above

   Answer: _____

8. What is the operation keyword(s) in the following question?

   What is three times $\dfrac{5}{12}$?

   Answer: _____

9. What is the math sentence for question 8?
   (a) $3 \times \dfrac{5}{12}$

   (b) $\dfrac{5}{12} \div 3$

   (c) $3 + \dfrac{5}{12}$

   (d) $3 \div \dfrac{5}{12}$

   Answer: _____

10. What is the operation keyword(s) in the following sentence?

    Double 22.
    (a) 22
    (b) Double
    (c) None of the above

    Answer: _____

11. 1 mug costs $1.50. Which math sentence will you use to find the cost of 15 mugs?
    (a) 1.50 + 15
    (b) 15 × 1.50
    (c) 15 ÷ 1.50
    (d) All of the above

    Answer: _____

1.6 Write a Math Expression with Multiple Operations (*)

Example 1:

What is the math sentence for the following expression?

8 multiplied by half of 20

(a) $\left(\dfrac{1}{2} \times 20\right) \div 8$

(b) $\left(\dfrac{1}{2} \div 20\right) + 8$

(c) $\left(\dfrac{1}{2} + 20\right) - 8$

(d) $\left(\dfrac{1}{2} \times 20\right) \times 8$

Solution:

We can write the math sentence as follows:

- half of 20 $= \dfrac{1}{2}$ of 20

  $= \dfrac{1}{2} \times 20$

- 8 multiplied by (half of 20)

  $= 8 \text{ multiplied by } \left(\dfrac{1}{2} \times 20\right)$

  $= \left(\dfrac{1}{2} \times 20\right) \times 8$

So the answer is (d).

Example 2:

What is the math sentence for the following expression?

5 added to the quotient of 15 and 3

(a) $(15 \times 3) + 5$
(b) $(15 \div 3) + 5$
(c) $(15 + 3) \div 5$
(d) $(15 \div 3) \times 5$

Solution:

We can write the math sentence as follows:

(5) added to (the quotient of 15 and 3)

- The quotient of 15 and 3 $= (15 \div 3)$

- (5) added to (the quotient of 15 & 3)
  $= $ (the quotient of 15 and 3) + (5)
  $= (15 \div 3) + 5$

So the answer is (b).

Choose the letter of the answer.

1. What is the math sentence for the following expression?

   $\frac{1}{4}$ added to the product of $\frac{1}{2}$ and $\frac{2}{3}$

   (a) $\left(\frac{1}{2}\times\frac{2}{3}\right)+\frac{1}{4}$

   (b) $\left(\frac{1}{2}\times\frac{2}{3}\right)-\frac{1}{4}$

   (c) $\left(\frac{1}{4}\times\frac{2}{3}\right)\times\frac{1}{2}$

   (d) $\left(\frac{1}{2}\times\frac{2}{3}\right)\div\frac{2}{3}$

   Answer: _____

2. What is the math sentence for the following expression?

   11 subtracted from the sum of 25 and 8

   (a) (25 × 8) + 11
   (b) (25 − 8) − 11
   (c) (25 + 8) − 11
   (d) (25 ÷ 11) + 8

   Answer: _____

3. What is the math sentence for the following expression?

   Triple of 12 added to 10

   (a) (12 × 3) − 10
   (b) (12 ÷ 3) × 10
   (c) (12 × 3) + 10
   (d) (12 × 3) ÷ 10

   Answer: _____

4. What is the math sentence for the following expression?

   7 less than 20% of 60

   (a) (0.2 × 60) − 7
   (b) (0.2 ÷ 60) + 7
   (c) (0.2 × 60) ÷ 7
   (d) (0.2 + 60) × 7

   Answer: _____

5. What is the math sentence for the following expression?

   Two-fifths of 100 added to 30

   (a) $\left(\frac{2}{5}\times100\right)\times30$

   (b) $\left(\frac{2}{5}\div100\right)-30$

   (c) $\left(\frac{2}{5}+100\right)\times30$

   (d) $\left(\frac{2}{5}\times100\right)+30$

   Answer: _____

6. What is the math sentence for the following expression?

   Product of $18.00 and $4.00 distributed equally among 8 people

   (a) (18 ÷ 8) − 4
   (b) (18 × 4) ÷ 8
   (c) (18 − 4) + 8
   (d) (18 + 8) ÷ 4

   Answer: _____

1.7 Review of Chapter 1 (*)

Write or choose the letter of the answer.

1. What operation will you use for the keywords *quotient of*?
   - (a) Division
   - (b) Addition
   - (c) Multiplication
   - (d) Subtraction

   Answer: _____

2. What is the operation keyword(s) in the following expression?

   Product of 10 and 5

   Answer: _____

3. What operation will you use for the keyword(s) in question 2?
   - (a) Multiplication
   - (b) Addition
   - (c) Division
   - (d) Subtraction

   Answer: _____

4. What is the math sentence for the following problem?

   Max had 12 pencils. He took 6 pencils from his brother. How many pencils did Max have in total?
   - (a) $12 \div 6$
   - (b) 12×6
   - (c) $12 - 6$
   - (d) $12 + 6$

   Answer: _____

5. What operation will you use for the keyword *times*?
   - (a) Subtraction
   - (b) Addition
   - (c) Multiplication
   - (d) Division

   Answer: _____

6. What operation will you use for the keywords *all together*?
   - (a) Multiplication
   - (b) Addition
   - (c) Division
   - (d) Subtraction

   Answer: _____

7. What is the math sentence for the following expression?

   Triple of 60 subtracted from 200
   - (a) $200 - (3 \times 60)$
   - (b) $200 - (3 + 60)$
   - (c) $200 + (60 \div 3)$
   - (d) $60 - (3 \times 200)$

   Answer: _____

8. What is the operation keyword(s) in the following sentence?

   Nil saves $8.00 every day.

   Answer: _____

Write or choose the letter of the answer.

9. What is the operation keyword(s) in the following sentence?

    Divide 55 by 11.

    Answer: _____

10. What operation will you use for the keyword(s) in question 9?
    (a) Subtraction
    (b) Division
    (c) Multiplication
    (d) Addition

    Answer: _____

11. Nikhil's daily wage is $7.00. He receives an increase of $3.00. What is his new wage?

    Which math sentence will you use to answer?

    (a) 7 – 3
    (b) 7 × 3
    (c) 7 + 3
    (d) All of the above

    Answer: _____

12. What operation will you use for the keyword *raise*?
    (a) Addition
    (b) Division
    (c) Subtraction
    (d) Multiplication

    Answer: _____

13. Lim bought 28 colors. He used 12 of them and gave 10 colors to his sister.

    Which math sentence can we use to find how many colors were left?
    (a) 28 – (12 + 10)
    (b) 28 + (12 + 10)
    (c) 28 – (12 – 10)
    (d) 12 × (28 + 10)

    Answer: _____

14. What is the operation keyword(s) in the following sentence?

    Arun had a remainder of $5.00 after shopping.

    Answer: _____

15. What is the math sentence for the following expression?

    two-thirds of 12

    (a) $\frac{2}{3} \times 12$

    (b) $\frac{2}{3} \div 12$

    (c) $12 + \frac{2}{3}$

    (d) All of the above

    Answer: _____

16. What is the operation keyword(s) in the following sentence?

    Multiply 20 and 25.

    Answer: _____

2. Basic Problem-Solving Strategies

2.1 One-Step Problems (*)

Example 1:

Bibek bought 4 pairs of trousers. He paid $48.00 for the trousers. What was the cost of 1 pair of trousers?

Solution:

The following information is given:

Number of trousers bought = 4

Total cost = $48.00

You can find the cost of 1 pair of trousers by dividing the total cost by the number of trousers.

(cost of 1 pair of trousers)

= (total cost)
    ÷ (number of trousers)

= $48.00 ÷ 4

= $12.00

So the cost of 1 pair of trousers was $12.00.

Example 2:

A house has 7 rooms. Each room has 2 windows. What operation will you use to find the total number of windows?

Solution:

Number of rooms in the house = 7

Number of windows in each room = 2

We can find the total number of windows by multiplying the number of rooms by the number of windows in each room.

So we can use multiplication to find the total number of windows.

Write or choose the letter of the answer.

1. A bag has 5 wallets. Each wallet has 6 card holders. What operation will you use to find the total number of card holders?
   (a) Multiplication
   (b) Subtraction
   (c) Division
   (d) Addition

   Answer: _____

2. What is the operation keyword(s) in the following problem?

   Kelvin has 10 pencils. Killy has 3 fewer pencils than Kelvin. How many pencils does Killy have?

   Answer: _____

Write or choose the letter of the answer.

3. Jay can read 10 pages in 1 hour. How many pages can he read in 3 hours?

    Answer: ____ _____
    unit

4. Raaj had 15 candies. He distributed them evenly among 3 friends. What operation will you use to find the number of candies each friend got?

    Answer: _____

5. If 1 basket can hold 10 pounds of fruit, how many baskets will hold 40 pounds of fruit?

    Answer: ____ _____
    unit

6. A tap took 45 minutes to fill a tank. How much time will it take if the tap has to fill two-fifths of the tank?

    Answer: ____ _____
    unit

7. Manoj can make 50 designs in 1 day. Anisa can make 62 designs in the same time. What is the total number of designs made by both of them in 1 day?

    Answer: ____ _____
    unit

8. Rahul bought 6 notebooks. If he spent $9.00 in total, what is the cost of each notebook?

    Answer: _____

9. What is the operation keyword(s) in the following problem?

    Nancy is 13 years old. Her brother is 2 times as old as Nancy.

    Answer: _____

10. What operation will you use for the keyword(s) in question 9?
    (a) Subtraction
    (b) Addition
    (c) Division
    (d) Multiplication

    Answer: _____

11. Mark has 7 crayons. His brother gave him 12 more crayons. What operation will you use to find the total number of crayons Mark has?

    Answer: _____

12. Mia bought 6 coffee mugs. She paid $18.00 for the coffee mugs. What was the cost of 1 coffee mug?

    Answer: _____

2.2 Multistep Problems 1 (**)

Example 1:

Georgia had 50 pounds of nuts in her shop. She sold one-half of the nuts and kept the rest of the nuts for her home.

If you want to find the number of nuts kept for her home, what question do you need to answer first?

    (a) How many nuts were there in her shop?
    (b) How many nuts did Georgia sell?
    (c) How many nuts did Georgia keep for her home?
    (d) All of the above

Solution:

You need to find the number of nuts sold before you can find the number of nuts kept for her home.

So the answer is (b).

Example 2:

How many nuts did Georgia sell in example 1?

Solution:

You can solve the problem as follows:

Number of nuts in the shop = 50 pounds

Number of nuts sold

$$= \frac{1}{2} \text{ of (number of nuts in shop)}$$

$$= \frac{1}{2} \text{ of } 50$$

$$= \frac{1}{2} \times 50 = 25 \text{ pounds}$$

So Georgia sold 25 pounds of nuts.

Example 3:

Jacob spent $80.00 in total. He spent half of the money to buy some dresses, $15.00 to buy a calculator, and the rest to buy sunglasses. How much money did he spend on sunglasses?

Solution:

The following information is given:

Total money spent = $80.00

Money spent on dresses
    = half of (total money spent)

$$= \frac{1}{2} \text{ of } \$80.00$$

Money spent on calculator = $15.00

You can use the following steps to find the answer:

Step 1: Find the amount of money spent on dresses.

Money spent on dresses

$$= \frac{1}{2} \text{ of } \$80.00$$

$$= \frac{1}{2} \times 80 = \frac{80}{2} = \$40.00$$

Step 2: Find the amount of money spent on the dresses and calculator.

Money spent on dresses and calculator
    = $40.00 + $15.00 = $55.00

Step 3: Find the amount of money spent on sunglasses.

Money spent on sunglasses
    = (total money spent) − (money spent on dresses and calculator)

    = $80.00 − $55.00

    = $25.00

So Jacob spent $25.00 on sunglasses.

Write or choose the letter of the answer.

1. Nelson has to make 36 paper crafts for his room. He has already made one-third of the paper crafts this morning. To find the remaining number of paper crafts, what question do you need to answer first?
   (a) How many paper crafts has Nelson already made?
   (b) How many paper crafts does he have to make in total?
   (c) How many paper crafts does he want to make?
   (d) All of the above

   Answer: _____

2. Fred spent $57.00 in total. He spent $11.00 on headphones, $15.00 on a shirt, and the rest on a watch. How much money did he spend on the watch?

   Answer: _____

3. A shopkeeper had 110 flowers in his shop. There were 50 roses, 30 lotuses, and the rest marigolds. How many marigolds were there?

   Answer: ____ _____
              unit

4. Mr. Hales bought some snacks. He paid $12.50 for popcorn and $4.50 for cakes. He was left with the same amount of money as he spent at the store. How much did he have at the beginning?

   Answer: _____

5. Mr. Rao has to use 50 pounds of sugar for his restaurant. He has already used three-fifth of the sugar.

   To find the remaining sugar, what question do you need to answer first?
   (a) How much sugar does Mr. Rao have to use in total?
   (b) How much sugar does he want to use?
   (c) How much sugar has he already used?
   (d) All of the above

   Answer: _____

6. Anisha had $28.00 to buy nuts. She spent one-half of the money on cashew nuts and the rest on almonds. How much money did Anisha spend on almonds?

   Answer: _____

2.3 Multistep Problems 2 (**)

Example 1:

Kile went to a fair with his friends. He bought 7 toys with a cost of $3.00 per toy. He also bought some home appliances worth $26.50. How much money did he spend in total?

Solution:

The following information is given:

Cost of 1 toy = $3.00

Number of toys bought = 7

Cost of home appliances = $26.50

You can use the following steps to find the answer:

Step 1: Find the cost of 7 toys.

Cost of 7 toys

= cost of 1 toy × 7

= $3.00 × 7 = $21.00

Step 2: Find the amount of money spent in total.

Total money spent

= cost of 7 toys
    + cost of home appliances

= $21.00 + $26.50 = $47.50

So Kile spent $47.50 in total.

Example 2:

The cost of 1 pizza is $6.50. Amit bought 6 pizzas and gave $40.00 to the cashier. How much money will the cashier return?

Solution:

The following information is given:

Cost of 1 pizza = $6.50

Number of pizzas bought = 6

Amount given to the cashier = $40.00

We can find the answer as given below.

Cost of 6 pizzas

= cost of 1 pizza
        × number of pizzas

= $6.50 × 6

= $39.00

Amount returned by the cashier

= amount given to the cashier
        − cost of 6 pizzas

= $40.00 − $39.00 = $1.00

So the cashier will return $1.00.

Write or choose the letter of the answer.

1. The cost of 1 book is $12.00. Nil bought 8 books for his shop. How much money will the cashier return if he gave $100.00 to the cashier?

Answer: _____

2. The cost of 1 carpet is $22.00. A vendor delivered 3 carpets to Jessi. How much money will the cashier return if she paid $70.00 for all the carpets?

Answer: _____

Write or choose the letter of the answer.

3. A pack of noodles costs $10.50. Andy bought 5 of them and gave $60.00 to the cashier.

   What question do you need to answer first to find how much money the cashier will return?
   - (a) What is the cost of 5 noodle packs?
   - (b) How many noodle packs did Andy buy?
   - (c) How much did 1 noodle pack cost?
   - (d) All of the above

   Answer: _____

4. Bill spent 7 hours preparing notes. He prepared English notes for 1.5 hours, Hindi notes for 3 hours, and the rest of the time prepared history notes. How much time did he spend preparing history notes?

   Answer: _____ _____

                           unit

5. Elina went to a mall with her friends. She bought 3 pairs of jeans with a cost of $18.00 per pair. She also bought some toys worth $15.00. How much money did she spend in total?

   Answer: _____

6. A coffee maker costs $25.50. Mrs. Wilson bought 4 coffee makers. What amount will the cashier return if she gave $110.00 to the cashier?

   Answer: _____

7. A belt costs $12.50. Mark bought 6 belts and gave $80.00 to the cashier.

   What question do you need to answer first to find how much money the cashier will return?
   - (a) How much did 1 belt cost?
   - (b) What is the cost of 6 belts?
   - (c) How many belts are there in the store?
   - (d) All of the above

   Answer: _____

8. What is the cost of 6 belts in question 7?

   Answer: _____

9. The cost of 1 iron is $24.25. Vinod delivered 8 irons to an apartment. How much money will Vinod return if he received $200.00 for all the irons?

   Answer: _____

2.4 Work Backward 1 (**)

Example 1:

Wilson bought 5 sandwiches for $1.50 each and 4 ice-cream cones for $2.50 each from a store. If the storekeeper returned $2.50, how much money did Wilson give to the storekeeper?

Solution:

The following information is given:

Cost of 1 sandwich = $1.50

Cost of 1 ice-cream cone = $2.50

Money returned by storekeeper = $2.50

We can use the following steps to find the answer.

- Find the cost of 5 sandwiches and 4 ice-cream cones.

  (cost of 5 sandwiches) = $1.50 × 5

  = $7.50

  (cost of 4 ice-cream cones)

  = $2.50 × 4 = $10.00

- Find the cost of all supplies.

  (cost of all supplies)

  = $7.50 + $10.00 = $17.50

- Find the amount given to the storekeeper.

  (amount given to the storekeeper)

  = $17.50 + $2.50 = $20.00

So Wilson gave $20.00 to the storekeeper.

Example 2:

Robin delivered 150 newspapers in total. He delivered an equal number of newspapers on 5 days of the week and delivered 35 newspapers during the weekend. How many newspapers did he deliver on each weekday?

Solution:

The following information is given:

Total newspapers delivered = 150

Number of newspapers delivered on the weekend = 35

Number of weekdays = 5

You can solve this problem by working backward.

Step 1: Find the number of newspapers delivered on weekdays.

(number of newspapers delivered on weekdays)

= (total newspapers delivered)
 − (newspapers delivered on weekend)

= 150 − 35 = 115 newspapers

Step 2: Find the number of newspapers delivered on each weekday.

(newspapers delivered on each weekday)

= (newspapers delivered on weekdays)
                    ÷ (number of weekdays)

= 115 ÷ 5 = 23 newspapers

So Robin delivered 23 newspapers on each weekday.

Write the answer.

1. Daniel ordered 3 cell phones for $30.00 each and 4 flash drives for $12.00 each from an online site. If the delivery boy returned $12.00, how much money did he pay the delivery boy?

   Answer: _____

2. Maria bought 3 pens for $0.50 each, 4 key rings for $0.25 each, and 1 geometry box for $1.25 from a store. If the cashier returned $1.25, how much money did she give to the cashier?

   Answer: _____

3. Jacob spent $46.00 at a store. He bought 5 pairs of pants that cost the same amount and a wallet for $11.00. What was the cost of each pair of pants?

   Answer: _____

4. A plant was 15 inches tall on Sunday. It had grown 4 inches from Thursday to Sunday. It had grown 5 inches from Monday to Thursday. How tall was the plant on Monday?

   Answer: ____ _____
   unit

5. Mr. Campbell went to buy office appliances. He paid a total of $250.00 for the appliances and got back $15.00 in change from the cashier. How much money did Mr. Campbell give to the cashier?

   Answer: _____

6. Some workers cleaned 80 rooms in total. They cleaned an equal number of rooms for 6 days of the week and cleaned 20 rooms during the weekend. How many rooms did they clean on each weekday?

   Answer: ____ _____
   unit

7. Ameli went to buy some personal expenses. She gave a total of $21.65 and got back $2.50 in change from the cashier. How much money did Ameli spend for her personal expenses?

   Answer: _____

8. Samuel and his friends spent $75.00 in a store. They bought 5 pairs of trousers that each cost the same amount. He and his friends also bought movie tickets for $15.00. What was the cost of each pair of trousers?

   Answer: _____

2.5 Work Backward 2 (**)

Example 1:

Mr. Harris bought some sweets for his daughter's birthday party. He gave $18.00 to the cashier, and the cashier returned $2.50. What was the cost of the sweets?

Solution:

The following information is given:

Money given to the cashier = $18.00
Money returned by the cashier = $2.50

You can solve this problem by the following step.

Cost of sweets
= (money given to the cashier)
  − (money returned by the cashier)
= $18.00 − $2.50 = $15.50

So the cost of sweets was $15.50.

Example 2:

After paying $25.25 for books, $12.50 for magazines, and $4.75 for pens, Mahi had $4.35 left. How much money did she have at the beginning?

Solution:

The following information is given:

Money paid for books = $25.25

Money paid for magazines = $12.50

Money paid for pens = $4.75

Money left = $2.50

We can find the amount of money she had at the beginning by adding all expenses and the amount left at the end.

Money at the beginning
= (money paid for books)
  + (money paid for magazines)
  + (money paid for pens)
  + (money left)

= $25.25 + $12.50 + $4.75 + $2.50

= $45.00

So Mahi had $45.00 at the beginning.

Write the answer.

1. The cost of one sweet packet is $7.50. Joy bought 8 packets for his home. How much money will the cashier return if he gave $64.00 to the cashier?

2. Nancy spent $33.00 at a store. She bought 4 bracelets that cost the same amount and some snacks for $5.00. What is the cost of each bracelet?

Answer: _____

Answer: _____

Write the answer.

3. The lunch recess at St. Joseph's School ended at 12:40 p.m. If the recess was for a period of 30 minutes, what time did the recess start?

Answer: _____ _____
    unit

4. After paying $24.00 for groceries, $8.50 for vegetables, and $4.75 for snacks, Mr. Paul had $3.50 left. How much money did he have at the beginning?

Answer: _____

5. Michael takes 45 minutes to spray paint a wall and 60 minutes to make wall art. He also takes a 15-minute break to reload supplies between jobs. If he finished everything at 4:00 p.m., what time did he start his work?

Answer: _____ _____
    unit

6. The cost of one snacks pack is $1.00. Manoj bought 45 packets for a small party in his home. How much money will the cashier return if Manoj gave $50.00 to the cashier?

Answer: _____

7. Edwin bought some crayons for his sister. He gave $4.50 to the cashier, and the cashier returned $0.25. What was the cost of the crayons?

Answer: _____

8. Mrs. Murphy spent $100.00 for a saree, $45.00 for bangles, and $25.00 for sandals for her friend's marriage party. She had $10.00 left. How much money did she have at the beginning?

Answer: _____

9. Richa went to buy kitchen appliances. She paid a total of $35.25 for the appliances and got back $4.75 in change from the cashier. How much money did Richa give to the cashier?

Answer: _____

10. After paying $15.00 for a backpack, $7.50 for notebooks, and $22.00 for a uniform set, Pamela had $3.50 left. How much money did she have at the beginning?

Answer: _____

2.6 Too Much and Too Little Information (*)

Example 1:

Review the question given below, and choose the best choice for the information available.

Joy has 25 backpacks in his store. He bought 5 more backpacks and 10 wallets. How many backpacks does he have?

(a) Too little information
(b) Too much information
(c) The right amount of information

Solution:

The following information is given:

Number of backpacks Joy has = 25
More backpacks Joy bought = 5
Number of wallets Joy bought = 10

You can find the total backpacks by adding the backpacks Joy had before and the number of backpacks he bought.

The information about wallets is not required to solve this problem. So we have too much information in this question.

So the answer is *too much information*, which is option (b).

Example 2:

Review the question given below, and choose the best choice for the information available.

There are 15 apples and 10 oranges in Basket 1. There are more apples and oranges in Basket 2. How many fruits are there in both the baskets?

(a) Too much information
(b) The right amount of information
(c) Too little information

Solution:

The following information is given:

Number of apples in Basket 1 = 15
Number of oranges in Basket 1 = 10

There are more apples and oranges in Basket 2, but the amount is not given.

Without knowing the amount of apples and oranges in Basket 2, we cannot find the total number of fruits in both the baskets.

So the answer is *too little information*, which is option (c).

Write or choose the letter of the answer.

1. If the following question has enough information, find the answer to the question. Otherwise, write *no answer* for the answer.

   There are 30 boys and 24 girls in grade 4. 12 boys from grade 4 left the school. How many students remained in grade 4?

   Answer: _____

2. Review the question given below, and choose the best choice for the information available.

   James needs to buy a storybook. How much money does he need to pay for the book?
   (a) Too much information
   (b) Too little information
   (c) The right amount of information

   Answer: _____

Write or choose the letter of the answer.

3. Review the question given below, and choose the best choice for the information available.

> Bill has 56 CDs in his music collection. He bought 10 more CDs, 5 pen drives, and a photo album. How many CDs does he have?

 (a) Too little information
 (b) Too much information
 (c) The right amount of information

Answer: _____

4. If the following question has enough information, find the answer to the question. Otherwise, write *no answer* for the answer.

> Andy wants to play baseball with his friends. How many total players are there on the team?

Answer: _____

5. Review the question given below, and choose the best choice for the information available.

> There are 36 deer and 19 monkeys in the San Francisco Zoo. There are more deer and monkeys in the Oakland Zoo. How many deer are there in both of the zoos?

 (a) Too much information
 (b) The right amount of information
 (c) Too little information

Answer: _____

6. Review the question given below, and choose the best choice for the information available.

> Bob has to solve 80 math problems in an hour. He has finished 58 problems. How many more problems does Bob need to solve?

 (a) The right amount of information
 (b) Too little information
 (c) Too much information

Answer: _____

7. Review the question given below, and choose the best choice for the information available.

> Sarika has 19 crayons and 6 pencils. She gives 8 crayons to her brother. How many crayons does she have left?

 (a) Too much information
 (b) The right amount of information
 (c) Too little information

Answer: _____

8. Review the question given below, and choose the best choice for the information available.

> There are 30 Nokia phones and 25 Samsung phones in a store. 10 Nokia phones are sold in a week. How many phones remain in the store?

 (a) Too little information
 (b) Too much information
 (c) The right amount of information

Answer: _____

2.7 Interpret the Remainder—1 (**)

Example 1:

Select the right choice for using the remainder to answer the following question.

Carlos needs 92 sweets for the Independence Day celebration. If each packet holds 10 sweets, how many packets will he need to buy?
(a) Drop the remainder and increase the quotient by 1
(b) Drop the remainder
(c) Use the remainder as the answer

Solution:

The following information is given:

Amount of sweets needed = 92 pieces
Number of sweets in 1 packet = 10

- First divide 92 by 10 to get the number of sweet packets.

  $92 \div 10 = 9 \text{ R } 2$

- $9 \times 10 = 90$
  that means 9 packets hold 90 sweets, but Carlos needs 92 sweets.

- Since 9 packets hold fewer than 92 sweets, increase the quotient by 1.

- Carlos needs to buy $9 + 1 = 10$ packets of sweets.

So the answer is *drop the remainder and increase the quotient by 1*, which is option (a).

Example 2:

Select the right choice for using the remainder to answer the following question.

If 140 candies were divided equally among several students, how many students will get 6 candies each?
(a) Use the remainder as the answer
(b) Ignore the remainder and use quotient as the answer
(c) Drop the remainder

Solution:

The following information is given:

Total candies = 140
Number of candies each student gets
= 6 candies

- First divide the total candies by the number of candies each student gets to find the number of students.

  $140 \div 6 = 23 \text{ R } 2$

- The remainder, 2, is not enough for another 6 candies. Therefore, drop the remainder.

- Only take 23 as the answer. 23 students will get 6 candies each.

So the answer is *drop the remainder*, which is option (c).

Write or choose the letter of the answer.

1. Select the right choice for using the remainder to answer the following question.

    Samuel has to pack 98 books in several boxes. If each box holds 10 books, how many boxes does he need?
    (a) Drop the remainder
    (b) Drop the remainder and increase the quotient by 1
    (c) Use the remainder as the answer

    Answer: _____

2. Jack had 46 marbles. He divided the marbles equally among 8 friends and kept the remaining marbles with him. How many marbles did Jack keep with him?

    Answer: _____ _____
    unit

3. Select the right choice for using the remainder to answer the following question.

    Max arranged 35 books in 4 bookshelves with an equal number of books in each bookshelf. How many books are left over?
    (a) Use the remainder as the answer
    (b) Drop the remainder
    (c) Drop the remainder and increase the quotient by 1

    Answer: _____

4. Select the right choice for using the remainder to answer the following question.

    137 apples were divided equally among several children. If each child gets 7 apples, how many children will get 7 apples each?
    (a) Drop the remainder and increase the quotient by 1
    (b) Use the remainder as the answer
    (c) Drop the remainder

    Answer: _____

5. Select the right choice for using the remainder to answer the following question.

    There are 87 stickers in a box. If 4 stickers are required for 1 poster, how many posters can be made with a box of stickers?
    (a) Drop the remainder and increase the quotient by 1
    (b) Drop the remainder
    (c) Use the remainder as the answer

    Answer: _____

6. Rohan baked 76 cookies for the school picnic. He divided the cookies into several packages with 6 cookies in each package. He gave the leftover cookies to his sister. How many cookies did he give to his sister?

    Answer: _____ _____
    unit

2.8 Interpret the Remainder—2 (**)

Example 1:

Select the right choice for using the remainder to answer the following question.

Simon had 80 baseball cards. He divided the cards among 6 friends so that each friend got an equal number of cards. He kept the remaining cards for himself. How many cards did he keep for himself?

(a) Drop the remainder and increase the quotient by 1
(b) Drop the remainder
(c) Use the remainder as the answer

Solution:

The following information is given:

Total number of baseball cards = 80
Number of friends = 6

- First divide the total number of baseball cards by the number of friends to find the number of cards for each friend.

$$80 ÷ 6 = 13 \text{ R } 2$$

- 6 friends will get 13 cards each, and 2 cards will be left over.

- Since he kept the remaining cards for himself, Simon kept 2 cards for himself.

- Only take the remainder 2 as the answer.

So the answer is *use the remainder as the answer,* which is option (c).

Example 2:

Lucy wants to plant 6 rows of rose plants with 10 plants in each row. The plants are available in bunches. If there are 5 plants in a bunch, how many bunches does Lucy need to buy?

Solution:

The following information is given:

Number of rows = 6
Number of plants in each row = 10
Number of plants in 1 bunch = 5

- First find the total number of plants by multiplying the number of rows by the number of plants in each row.

$$\text{Total number of plants} = 6 × 10$$
$$= 60 \text{ plants}$$

- Then find the number of bunches by dividing total number of plants by the number of plants in 1 bunch.

$$\text{Number of bunches needed}$$
$$= 60 ÷ 5 = 12 \text{ bunches}$$

So Lucy needs to buy 12 bunches of plants.

Write or choose the letter of the answer.

1. Select the right choice for using the remainder to answer the following question.

   Sivani needs to distribute 46 ice-cream cones among 20 children. If each child gets an equal number of cones, how many ice-cream cones will each child get?

   (a) Drop the remainder
   (b) Use the remainder as the answer
   (c) Drop the remainder and increase the quotient by 1

   Answer: _____

2. There were 75 candies in a box. Mr. Woods divided the candies equally among 8 children and kept the remaining candies with him. How many candies did Mr. Woods keep with him?

   Answer: ____ _____
   　　　　　　　unit

3. Select the right choice for using the remainder to answer the following question.

   Jasmin plucked 38 rose flowers from her garden. She plucked an equal number of red roses and white roses. How many white roses did she pluck?

   (a) Drop the remainder and increase the quotient by 1
   (b) Use the remainder as the answer
   (c) Drop the remainder

   Answer: _____

4. Kile has to write 38 pages in a week. He divided the pages into several days with 5 pages in each day. He gave the remaining pages to his sister to write for him. How many pages did his sister write for him?

   Answer: ____ _____
   　　　　　　　unit

5. Select the right choice for using the remainder to answer the following question.

   Karan has to pack 46 bottles. If each box holds 6 bottles, how many boxes does Karan need?

   (a) Drop the remainder and increase the quotient by 1
   (b) Use the remainder as the answer
   (c) Drop the remainder

   Answer: _____

6. Select the right choice for using the remainder to answer the following question.

   Peter has 70 photos. He gave 8 photos to each student in the school. How many students will get 8 photos each?

   (a) Use the remainder as the answer
   (b) Drop the remainder
   (c) Drop the remainder and increase the quotient by 1

   Answer: _____

2.9 Review of Chapter 2 (**)

Write or choose the letter of the answer.

1. Jack bought some snacks. He paid $13.00 for cookies and $5.00 for chips. He was left with the same amount of money as he spent at the store. How much did he have at the beginning?

   Answer: _____

2. Mr. Ray has to sell 40 shirts in his shop. He has already sold 1/4 of the shirts.

   To find the remaining number of shirts, what question do you need to answer first?
   (a) How many shirts has he already sold?
   (b) How many shirts does Mr. Ray have to sell in total?
   (c) How many shirts does he want to sell?
   (d) All of the above

   Answer: _____

3. Mr. Harper bought some ice-cream cones for a small party in his home. He gave $50.00 to the cashier, and the cashier returned $3.50. What was the cost of the ice-cream cones?

   Answer: _____

4. Rahul bought 5 chairs. He paid $125.00 for the chairs. What was the cost of 1 chair?

   Answer: _____

5. Review the question given below and choose the best choice for the information available.

   Franc has 7 pens with him. He bought 5 more pens and 7 pencils. How many pens does he have?
   (a) Too much information
   (b) Too little information
   (c) The right amount of information

   Answer: _____

6. Angela spent $76.00 in total. She spent half of the money to buy a phone, $15.00 to buy a makeup kit, and the rest to buy sandals. How much money did she spend on sandals?

   Answer: _____

7. A workbook has 10 pages. Each page has 8 practice questions. What operation will you use to find the total number of questions?

   Answer: _____

Write or choose the letter of the answer.

8. If the following question has enough information, find the answer to the question. Otherwise, write *no answer* for the answer.

    There are 20 cows and 25 goats in a field. 15 of the cows leave. How many of them remain in the field?

    Answer: ____ _____
                            unit

9. Mark went to watch a movie with his friends. He bought 3 movie tickets with a cost of $2.00 per ticket. He also bought some snacks worth $5.50. How much money did he spend in total?

    Answer: _____

10. Select the right choice for using the remainder to answer the following question.

    Terry has to organize 75 chairs in a room. If he can put a maximum of 8 chairs at each table, how many tables in the room will have 8 chairs each?

    (a) Drop the remainder and increase the quotient by 1
    (b) Use the remainder as the answer
    (c) Drop the remainder

    Answer: _____

11. Review the question given below and choose the best choice for the information available.

    Eli wanted to buy 5 pumpkins from a store. Each pumpkin cost $3. If she gave $20 to the cashier, how much change did she get back?

    (a) Too much information
    (b) Too little information
    (c) The right amount of information

    Answer: _____

12. Daniel bought 4 pizzas for $3.50 each and 5 burgers for $3.00 each from a store. If the storekeeper returned $3.50, how much money did Daniel give to the storekeeper?

    Answer: _____

13. The cost of 1 novel is $3.00. Arnav bought 3 novels and gave $10.00 to the cashier. How much money did the cashier return?

    Answer: _____

14. There are 36 muffins in a pack. Mr. Woods divided the muffins equally among 5 children and kept the remaining muffins with him. How many muffins did Mr. Woods keep with him?

    Answer: ____ _____
                            unit

3. Unitary Method

3.1 Unitary Method—Direct Proportion (*)

Example 1:

If 3 bottles can carry 6 liters of water, how much water will 1 bottle carry?

Solution:

We can solve this problem as follows:

Amount of water 3 bottles can carry = 6 l

Amount of water 1 bottle will carry
$$= 6 \div 3$$
$$= 2 \text{ liters}$$

So 1 bottle will carry 2 liters of water.

Note:

More bottles will carry more water.

Fewer bottles will carry less water.

Example 2:

If 2 kids can eat 6 muffins, how many muffins can 6 kids eat?

Solution:

This problem can be solved using the following steps.

Step 1: Find the number of muffins 1 kid can eat.

Number of muffins 2 kids can eat = 6

Number of muffins 1 kid can eat
$$= 6 \div 2$$
$$= 3 \text{ muffins}$$

Step 2: Find the number of muffins 6 kids can eat.

Number of muffins 1 kid can eat = 3

Number of muffins 6 kids can eat
$$= 3 \times 6$$
$$= 18 \text{ muffins}$$

So 6 kids can eat 18 muffins.

Write the answer.

1. Joy walks 3 miles in 1 hour; how much distance will he cover in 2 hours?

    Answer: ____ _____
    unit

2. If 2 trucks can carry 50 tons at once, how much can 1 truck carry at once?

    Answer: ____ _____
    unit

Write the answer.

3. If 1 bowl can hold 3 fish, how many fish can be held in 4 bowls?

Answer: ____ _____
                    unit

4. If 3 students can develop 9 worksheets, how many worksheets can 12 students develop?

Answer: ____ _____
                    unit

5. If 2 people can paint 4 walls in a day, how many walls will 1 person paint?

Answer: ____ _____
                    unit

6. If 2 workers can clean 6 rooms in a day, how many rooms can 10 workers clean in a day?

Answer: ____ _____
                    unit

7. If 3 birds can make 1 nest in a day, how many birds can make 7 nests in a day?

Answer: ____ _____
                    unit

8. If 1 person can make 15 toys in a day, how many toys will 8 people make in a day?

Answer: ____ _____
                    unit

9. If 24 workers can dig 8 holes, how many workers will it take to dig 1 hole?

Answer: ____ _____
                    unit

10. If 1 rabbit can drink 13 liters of water in a week, how many rabbits will drink 52 liters of water in a week?

Answer: ____ _____
                    unit

11. If 3 tailors can stitch 6 shirts, how many shirts can 1 tailor stitch?

Answer: ____ _____
                    unit

12. If 4 children can eat 8 ice-cream cones, how many ice-cream cones can 10 children eat?

Answer: ____ _____
                    unit

3.2 Unitary Method—Inverse Proportion (*)

Example 1:

   2 students take 3 hours to complete an assignment. How many hours will it take 1 student to complete the same assignment?

Solution:

We can solve this problem as follows:

- Number of hours for 2 students to complete an assignment = 3 hours

- Number of hours for 1 student to complete the assignment = 3 × 2

       = 6 hours

So 1 student will take 6 hours to complete the same assignment.

Note:

   More students will complete the assignment in less time.

   Fewer students will complete the assignment in more time.

Example 2:

   If 1 person takes 15 days to finish a job, how many days will it take 5 people to finish the same job?

Solution:

We can solve this problem as follows:

- Number of days for 1 person to finish the job = 15 days

- Number of days for 5 people to finish the job = 15 ÷ 5 = 3 days

So 5 people will take 3 days to finish the job.

Note:

   Fewer people will finish a job in more time.

   More people will finish a job in less time.

Write the answer.

1. If 3 people take 6 days to make a brick wall, how many days will it take 1 person to make the same brick wall?

   Answer: ____ _____
                         unit

2. If 1 worker takes 6 days to mow a lawn, how many days will it take 2 workers to mow the same lawn?

   Answer: ____ _____
                         unit

3. 6 people can water the plants in 4 hours. 3 people could not come one day. How long did it take the other people to water the plants on that day?

   Answer: ____ _____
                         unit

4. If 1 kid takes 16 minutes to eat a candy packet, how much time will it take 4 kids to eat the same candy packet?

   Answer: ____ _____
                         unit

Write the answer.

5. 20 workers can repair a building in 12 days. How many days will 30 workers take to repair the building?

Answer: _____ _____
unit

6. If 1 person takes 9 hours to make special wall art, how many hours will it take 3 people to make the same art?

Answer: _____ _____
unit

7. 5 engineers can design a model of a car in 4 hours. How many hours will 2 engineers take to design the model?

Answer: _____ _____
unit

8. If 3 tailors take 5 days to make some uniforms, how many days will it take 1 tailor to make those uniforms?

Answer: _____ _____
unit

9. 5 machines can drill some holes in 12 minutes. How long will 3 machines take to drill the holes?

Answer: _____ _____
unit

10. If 1 machine takes 20 minutes to fill soft drinks in 100 bottles, how long will it take 4 machines to fill those bottles?

Answer: _____ _____
unit

11. If 5 children can complete a task in 3 hours, how long will it take 1 person to complete the same task?

Answer: _____ _____
unit

12. If 1 person takes 18 hours to write a manuscript, how many hours will it take 2 people to write the manuscript?

Answer: _____ _____
unit

13. 5 girls take 8 hours to make some dolls. How long will 10 girls take to make those dolls?

Answer: _____ _____
unit

14. 5 robots can assemble a bike in 4 hours. 1 robot could not work. How long did it take the other robots to assemble the bike?

Answer: _____ _____
unit

3.3 Unitary Method—Time Problems (**)

Example 1:

It takes 1 worker 8 hours to mow a lawn. How many hours will 4 workers take to mow the same lawn?

Solution:

You can solve this problem as follows:

Number of hours taken by 1 worker to mow a lawn = 8

Number of hours taken by 4 workers to mow the lawn = 8 ÷ 4

= 2 hours

So 4 workers will take 2 hours to mow the lawn.

Example 2:

Rubi and her brother take 3 days to make some paper crafts. How long will it take Rubi if she wants to make those paper crafts alone?

Solution:

You can solve this problem as follows:

Time taken by 2 people to make some paper crafts = 3 days

Time taken by 1 person to make those paper crafts = 3 × 2 = 6 days

So it will take 6 days if Rubi wants to make those paper crafts without her brother.

Example 3:

David takes 20 minutes to fill 10 water bottles. How long will he take to fill 15 water bottles?

Solution:

This problem can be solved using the following steps.

Step 1: Find the time taken to fill 1 bottle.

Time taken to fill 10 bottles = 20 minutes
Time taken to fill 1 bottle = 20 ÷ 10

= 2 minutes

Step 2: Find the time taken to fill 15 bottles.

Time taken to fill 1 bottle = 2 minutes
Time taken to fill 15 bottles = 2 × 15

= 30 minutes

So David will take 30 minutes to fill 15 water bottles.

Note:

Filling more bottles will take more time.

Filling fewer bottles will take less time.

Write the answer.

1. Amit takes 24 minutes to eat 12 bananas. How long will he take to eat 20 bananas?

   Answer: ____ _____
   <u>unit</u>

2. Angela and 2 of her sisters take 3 hours to type a seminar report. How long will it take Angela if she wants to type the report alone?

   Answer: ____ _____
   <u>unit</u>

3. A car takes 30 minutes to cover a 25-kilometer distance. How long will it take to cover 20 kilometers?

   Answer: ____ _____
   <u>unit</u>

4. Mrs. Wagner takes 5 hours to teach 4 subjects. How long will she take to teach 8 subjects?

   Answer: ____ _____
   <u>unit</u>

5. It takes 1 worker 40 minutes to clean a floor. How long will 2 workers take to clean the floor?

   Answer: ____ _____
   <u>unit</u>

6. Mia and 2 of her friends take 14 days to make a project. How long will it take Mia if she wants to make the project alone?

   Answer: ____ _____
   <u>unit</u>

7. A machine takes 1 hour to wrap 50 gifts. How long will it take to wrap 100 gifts?

   Answer: ____ _____
   <u>unit</u>

8. It takes 4 people 3 hours to decorate a marriage hall. How many hours will 1 person take to decorate the hall?

   Answer: ____ _____
   <u>unit</u>

9. It takes 1 child 45 minutes to eat a packet of sweets. How many minutes will it take 3 children to eat the sweets?

   Answer: ____ _____
   <u>unit</u>

10. Bill and his sister take 5 days to eat a bag of nuts. How long will it take Bill if he wants to eat the nuts alone?

    Answer: ____ _____
    <u>unit</u>

3.4 Unitary Method—Work Problems (**)

Example 1:

Nelson can complete a whole task in 10 minutes. What fraction of the task can he complete in 1 minute?

Solution:

Consider the whole task as 1 unit.

Task completed in 10 minutes = 1

Task completed in 1 minute

$$= (1 \div 10) \text{ units} \leftarrow \text{ divide by 10}$$

$$= \frac{1}{10} \text{ units}$$

$$= \frac{1}{10} \text{ of the whole task}$$

So Nelson can complete $\left(\dfrac{1}{10}\right)$ of the

whole task in 1 minute.

Example 2:

Emily wrote 35 pages of an assignment in 5 days. How many pages of the assignment will she write in 7 days?

Solution:

This problem can be solved using the following steps:

Step 1: Find the number of pages that can be written in 1 day.

Number of pages written in 5 days = 35
Number of pages written in 1 day = 35 ÷ 5
$$= 7 \text{ pages}$$

Step 2: Find the number of pages that can be written in 7 days.

Number of pages written in 1 day = 7
Number of pages written in 7 days = 7 × 7
$$= 49 \text{ pages}$$

So Emily will write 49 pages of the assignment in 7 days.

Write the answer.

1. Rob washed 15 cloths in 3 hours. How many clothes will he wash in 2 hours?

    Answer: _____ _____
                            unit

2. Amit can decorate his room in 5 hours. What fraction of the room can he decorate in 1 hour?

    Answer: _____

3. Sarah can design a painting in 40 minutes. What fraction of the painting can she do in 20 minutes?

    Answer: _____

4. A microwave oven can bake a cake in 25 minutes. What fraction of the cake can the oven bake in 1 minute?

    Answer: _____

Write the answer.

5. Michael runs 8 miles in 2 hours. How far will he run in 3 hours?

    Answer: _____ _____
    _____unit

6. Ronak can cook a dish in 20 minutes. What fraction of the dish can he cook in 5 minutes?

    Answer: _____

7. A worker can mow a field in 3 days. What fraction of the field can he mow in 1 day?

    Answer: _____

8. A tap can fill $\frac{1}{4}$ of a tank in 20 minutes. What fraction of the tank can it fill in 40 minutes? Write your answer as a fraction in simplest form.

    Answer: _____

9. Rahul eats 6 burgers in 2 hours. How many burgers will he eat in 5 hours?

    Answer: _____ _____
    _____unit

10. Sarah can do a nail paint in 15 minutes. What fraction of the nail paint can she do in 10 minutes?

    Answer: _____

11. A cow can produce 42 liters of milk in 7 days. How much milk can the cow produce in 5 days?

    Answer: _____ _____
    _____unit

12. Rakesh can paint $\frac{1}{3}$ of a hall in 2 hours. What fraction of the hall can he paint in 4 hours? Write your answer as a fraction in simplest form.

    Answer: _____

13. Franc can make 50 ice-cream cones in 2 hours. How many ice-cream cones can he make in 5 hours?

    Answer: _____ _____
    _____unit

14. A mouse can eat a slice of butter in 4 days. What fraction of the slice of butter can the mouse eat in 1 day?

    Answer: _____

3.5 Review of Chapter 3 (**)

Write the answer.

1. It takes 1 worker 6 hours to clean a garden. How many hours will 3 workers take to clean the same garden?

   Answer: ____ _____
                  unit

2. Mahi and her sister take 4 hours to make a bundle of handkerchiefs. How long will it take Mahi if she wants to make those handkerchiefs alone?

   Answer: ____ _____
                  unit

3. Rohit can read a novel in 5 days. What fraction of the novel can he read in 1 day?

   Answer: _____

4. If 10 students can develop 40 worksheets, how many students will it take to develop 60 worksheets?

   Answer: ____ _____
                  unit

5. James takes 30 minutes to arrange books in 5 shelves. How long will he take to arrange books in 8 shelves?

   Answer: ____ _____
                  unit

6. If 4 women can water 80 plants in a lawn, find the number of plants 10 women can water.

   Answer: ____ _____
                  unit

7. If 5 children can eat 20 cookies, how many cookies can 12 children eat?

   Answer: ____ _____
                  unit

8. If 1 person takes 8 hours to make a design, how many hours will it take 2 people to make the same design?

   Answer: ____ _____
                  unit

9. 5 engineers take 8 days to complete a project. How many days will it take 1 engineer to complete the same project?

   Answer: ____ _____
                  unit

10. If 5 friends can make 40 toys, how many toys can 3 friends make?

    Answer: ____ _____
                   unit

Write the answer.

11. If 2 cans hold 30 liters of milk, how much milk will 1 can hold?

Answer: ____ _____
                            unit

12. It takes 1 worker 50 minutes to dig a hole. How many minutes will 5 workers take to dig the same hole?

Answer: ____ _____
                            unit

13. If 1 person takes 18 days to finish a job, how many days will it take 6 people to finish the same job?

Answer: ____ _____
                            unit

14. Nikita and 2 friends take 2 hours to arrange all the chairs at the annual function party. How long will it take Nikita to arrange the chairs by herself?

Answer: ____ _____
                            unit

15. If 5 deer drink 20 liters of water per day, how many deer will drink 60 liters of water per day?

Answer: ____ _____
                            unit

16. If 7 people can make an interior in 4 days, how many days will it take 1 person to make the interior?

Answer: ____ _____
                            unit

17. Nancy can finish a task in 6 hours. How long will she take to finish $\frac{2}{3}$ of the task?

Answer: ____ _____
                            unit

18. If Nikhil eats 18 sweets in 2 hours, how long will he take to eat 45 sweets?

Answer: ____ _____
                            unit

19. If 5 jars can hold 75 marbles, how many marbles will 1 jar hold?

Answer: ____ _____
                            unit

20. Mark and his brother take 7 days to eat a jar of pickles. How long will it take Mark if he wants to eat the pickles alone?

Answer: ____ _____
                            unit

4. Number Problems

4.1 Place-Value Concepts (*)

Example 1:

What is the sum of the *place values* of 7 and 3 in 7,213?

Solution:

In the number 7,213:

Place value of 7 = 1,000
Place value of 3 = 1

Sum of the place values of 7 and 3
$$= 1,000 + 1$$
$$= 1,001$$

So the sum of the place values of 7 and 3 is 1,001.

Example 2:

I am the largest 2-digit number with 7 as my ones digit. What number am I?

Solution:

The largest 2-digit number is 99.

Ones digit = 7   ← given
Tens digit = 9

So the number is 97.

Example 3:

If you replace the tens digit by the double of the hundreds digit in the number 3,217, what is the new number?

Solution:

In the number 3,217:

Tens digit = 1
Hundreds digit = 2
Double of the hundreds digit = 2 × 2 = 4

If we replace the tens digit by the double of the hundreds digit in the number 3,217, then the tens digit will be 4 instead of 1.

So the new number is 3,247.

Example 4:

I am 400 more than the smallest possible 4-digit number that can be made using the digits 1, 2, 5, and 3. What number am I?

Solution:

The smallest 4-digit number is 1,000.

The smallest possible 4-digit number using the digits 1, 2, 5, and 3 is 1,235.

The number is 400 more than the smallest possible 4-digit number, which is 1,235.

$$= 1,235 + 400 = 1,635$$

So the number is 1,635.

Write the answer.

1. I am the smallest 3-digit number with 1 as my tens digit. What number am I?

    Answer: _____

2. What is the smallest 3-digit number without a leading zero?

    Answer: _____

3. What is the sum of the place values of 2, 6, and 4 in 52,614?

    Answer: _____

4. What number is twice the value of the hundred-thousands digit in 2,431,760?

    Answer: _____

5. I am the smallest 2-digit number with 2 as my ones digit. What number am I?

    Answer: _____

6. What is the largest 3-digit number with a leading zero?

    Answer: _____

7. I am 200 less than the largest possible 3-digit number that can be made using the digits 7, 9, and 8. What number am I?

    Answer: _____

8. If you replace the hundreds digit by the triple of the ones digit in the number 543, what is the new number?

    Answer: _____

9. What is the sum of the place values of 9 and 4 in 9,458?

    Answer: _____

10. I am the largest 3-digit number with 8 as my tens digit. What number am I?

    Answer: _____

11. What is the smallest 2-digit number without a leading zero?

    Answer: _____

12. What number is half of the value of the thousands digit in 6,837?

    Answer: _____

13. If you replace the thousands digit by the sum of the ones and tens digits in the number 2,714, what is the new number?

    Answer: _____

14. What is the sum of the place values of 1, 3, and 2 in 13,629?

    Answer: _____

4.2 Decimal Place-Value Concepts (**)

Example 1:

What is the *smallest* number greater than 8 with *one decimal place*?

Solution:

We can find the answer by using a decimal place-value table.

ones	decimal point	tenths

- The number has one decimal place. So place the decimal point in the second box from the right.

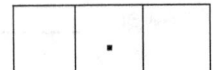

- The number is the smallest number that is greater than 8. This means we have 8 before the decimal point. Any other number before the decimal point will not be the smallest. So write 8 before the decimal point.

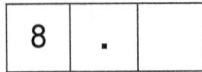

- For the number to be the smallest, the tenths digit has to be 1. If you write 0 at the tenths place, the number will not have any decimal place. Try other numbers for the digit, and check if the number is the smallest.

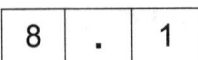

So the decimal number is 8.1.

Example 2:

I am the largest decimal number that is less than 9 with one decimal place having 6 in the tenths place. What number am I?

Solution:

We can find the answer by using a decimal place-value table.

ones	decimal point	tenths

- The number has one decimal place. So place the decimal point in the second box from the right.

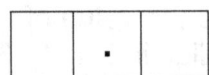

- The number is the largest number that is less than 9. This means we have 8 before the decimal point. Any other number before the decimal point will not be the largest. So write 8 before the decimal point.

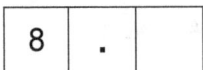

- The decimal place has 6 in the tenths place. So write 6 in the tenths place.

8	.	6

So the decimal number is 8.6.

Write the answer.

1. What is the sum of the values of 5 and 3 in 45.32?

Answer: _____

2. I am a number with 2 decimal places. My left-most digit is 2, and every digit is 2 times the digit to its left. What number am I?

Answer: _____

3. What is the smallest number with 3 decimal places that is greater than 32?

Answer: _____

4. I am the largest decimal number that is less than 15 with 2 decimal places having the same digit, 4. What number am I?

Answer: _____

5. What is the largest number less than 5 with 1 decimal place?

Answer: _____

6. I am a number with 2 decimal places. My right-most digit is 1, and every digit is 3 times the digit to its right. What number am I?

Answer: _____

7. I am the smallest decimal number that is greater than 22 with 1 decimal place having 8 in the tenths place. What number am I?

Answer: _____

8. What is the smallest number greater than 10 with 2 decimal places?

Answer: _____

9. What is the sum of the values of 1, 9, and 4 in 19.647?

Answer: _____

10. What is the largest number less than 9 with 3 decimal places?

Answer: _____

4.3 Rounding Numbers up to Billions (***)

Example 1:

Round 2,526,492 to the nearest million.

Solution:

To round a number:

- Find the place you want to round. Look 1 place to the right.
- If the digit is less than 5, keep the rounding digit the same (round down).
- If the digit is 5 or higher, increase the rounding digit by 1 (round up).
- Replace all the digits to the right of the rounding digit by 0.

We can find the answer by using the following steps.

Step 1: Find the digit in the million place. This is the place we want to round.
2,526,492

Step 2: Look 1 place to the right.
2, **5**26,492
This digit (5) is **equal to** 5.
So we need to **round up**.

Step 3: Round the number.
Replace all the digits to the right of the millions place by 0
2,526,492 → 2,**000,000**

To round up to nearest million, add 1 to the other part of the number (2+1=3).
2,000,000 → **3**,000,000

So 2,526,492 rounded to nearest million is 3,000,000.

Example 2:

I am the smallest number that never repeats a digit. When rounded to the nearest ten thousand, I round to 40,000. What number am I?

Solution:

We can find the answer by using a place-value table.

ten thousands	thousands	hundreds	tens	ones

- The number is the smallest number, which has to be 35,000. When rounded to the nearest ten thousand, it will round to 40,000. So write 35,000 in the place-value table.

3	5	0	0	0

- In this number we want to round the ten-thousands digit, which is **3**. The right place digit of 3 is **5**. So we need to round up. Replace all the digits to the right of the ten-thousands place by **0**, which is **30,000**. To round up the nearest ten thousand, add 1 to the ten-thousands digit (3 + 1 = 4). So the number is 40,000.

4	0	0	0	0

- We need to take the smallest number that never repeats a digit. So the number we can take 35,012.

3	5	0	1	2

So the number is 35,012.

Write the answer.

1. Round 5,764,215 to the nearest ten thousand.

    Answer: _____

2. I am the largest number that never repeats a digit. When rounded to the nearest thousand, I round to 25,000. What number am I?

    Answer: _____

3. When rounding to the nearest ten thousand, what is the smallest number that rounds to 40,000?

    Answer: _____

4. Round 584,397,136 to the nearest billion.

    Answer: _____

5. Jay solved 45 math and 28 English questions in an hour. About how many tens of questions did he solve?

    Answer: _____

6. Round 4,982 to the nearest hundred.

    Answer: _____

7. Ben and Adam participated in a race. Ben ran 5,200 meters. Adam ran 6,450 meters. About how many kilometers did they run in total?

    Answer: _____

8. When rounding to the nearest thousand, what is the smallest number that rounds to 3,500?

    Answer: _____

9. Round 75,479,326 to the nearest ten million.

    Answer: _____

10. I am the smallest number that never repeats a digit. When rounded to the nearest thousand, I round to 5,000. What number am I?

    Answer: _____

11. Nikita wrote 1,155 English words and 820 Hindi words in a day. About how many thousands of words did she write?

    Answer: _____

4.4 Different Forms to Write Numbers (*)

Example 1:

Write 216,258 in expanded form, and find the missing number in the following math sentence.

216,258 = 200,000 + _____ + 6,000
+ 200 + 50 + 8

Solution:

Write the numbers in the place value table.

hundred thousands	ten thousands	thousands	hundreds	tens	ones
2	1	6	2	5	8

We can write the number 216,258 in expanded form: 2 hundred thousands, 1 ten thousand, 6 thousands, 2 hundreds, 5 tens, and 8 ones.

2 hundred thousands = 200,000

1 ten thousand = 10,000

6 thousands = 6,000

2 hundreds = 200

5 tens = 50

8 ones = 8

216,258 = 200,000 + _10,000_ + 6,000
+ 200 + 50 + 8

So the missing number is 10,000.

Example 2:

I am given a number that can be written in expanded form as given below.

60,000 + 5,000 + 400 + 30 + 2

If I change 400 to 600 and 2 to 7, what will be the new number in standard form?

Solution:

First we can write each number in the expanded form:

60,000 = 6 ten thousands

5,000 = 5 thousands

400 = 4 hundreds

30 = 3 tens

2 = 2 ones

Then write the numbers in the place value table.

ten thousands	thousands	hundreds	tens	ones
6	5	4	3	2

60,000 + 5,000 + 400 + 30 + 2 = 65,432

If I change 400 to 600 and 2 to 7, the new number will be

60,000 + 5,000 + 600 + 30 + 7 = 65,637.

So the new number in standard form will be 65,637.

Write the answer.

1. Write 8,645,293 in expanded form, and find the missing number in the following math sentence.

    8,645,293 = _____ + 600,000

    + 40,000 + 5,000 + 200 + 90 + 3

    Answer: _____

2. I am given a number that can be written in expanded form as given below.

    7,000 + 200 + 60 + 8

    If I change 7,000 to 9,000, what will be the new number in standard form?

    Answer: _____

3. Select the answer that shows the word form of 542,350.

    (a) Five hundred fifty-two hundred three fifty hundred
    (b) Five hundred forty-two thousand three hundred and fifty
    (c) Five thousand forty-two hundred
    (d) None of the above

    Answer: _____

4. Write 52,370 in expanded form, and find the missing number in the following math sentence.

    52,370 = 50,000 + 2,000 + ___ + 70 + 0

    Answer: _____

5. I am given a number that can be written in expanded form as given below.

    400,000 + 70,000 + 500 + 80 + 4

    If I change 70,000 to 40,000 and 80 to 30, what will be the new number in standard form?

    Answer: _____

6. Select the answer that shows the word form of 61,805.

    (a) Six thousand one hundred eight hundred five
    (b) Six hundred one thousand and eight hundred five
    (c) Sixty-one thousand eight hundred and five
    (d) None of the above

    Answer: _____

7. Write 457,952 in expanded form, and find the missing number in the following math sentence.

    457,952 = 400,000 + 50,000 + 7,000

    + 900 + _____ + 2

    Answer: _____

8. Write the expanded form given below in standard form.

    9,000,000 + 70,000 + 4,000 + 300 + 8

    Answer: _____

4.5 Review of Chapter 4 (***)

Write the answers.

1. What is the largest 4-digit number with a leading zero?

   Answer: _____

2. What is the sum of the place values of 1, 9, and 4 in 18,934?

   Answer: _____

3. Write the expanded form given below in standard form.

   300,000 + 60,000 + 5,000 + 40 + 5

   Answer: _____

4. What is the smallest number with 2 decimal places that is greater than 10?

   Answer: _____

5. What is the sum of the values of 2 and 8 in 24.18?

   Answer: _____

6. I am the smallest number that never repeats a digit. When rounded to the nearest thousand, I round to 9,000. What number am I?

   Answer: _____

7. Round 78,254,193 to the nearest ten million.

   Answer: _____

8. What is the smallest 3-digit number without a leading zero?

   Answer: _____

9. I am 500 more than the smallest possible 4-digit number that can be made using the digits 5, 1, 3, and 2. What number am I?

   Answer: _____

10. What is the largest number with 3 decimal places that is less than 25?

    Answer: _____

11. Write 12,570 in expanded form, and find the missing number in the following math sentence.

    12,570 = 10,000 + _____ + 500 + 70

    Answer: _____

Write the answers.

12. What is the smallest number with 3 decimal places that is greater than 8?

Answer: _____

13. I am a number with 3 decimal places. My left-most digit is 1 and every digit is 2 times the digit to its left. What number am I?

Answer: _____

14. What is the difference between the place values of 3 and 8 in 435.82?

Answer: _____

15. What is the smallest number greater than 88 with 1 decimal place?

Answer: _____

16. Write 12,037,485 in expanded form, and find the missing number in the following math sentence.

12,037,485 = _____ + 2,000,000
+ 30,000 + 7,000 + 400 + 80 + 5

Answer: _____

17. If you replace the hundreds digit by triple the ones digit in the number 8,542, what is the new number?

Answer: _____

18. I am given a number that can be written in expanded form as given below.

600,000 + 50,000 + 7,000 + 50 + 6

If I change 50,000 to 20,000 and 50 to 90, what will be the new number in standard form?

Answer: _____

19. What is the difference between the values of 7 and 8 in 721.83?

Answer: _____

20. Bob had 225 red marbles and 300 white marbles with him. About how many hundreds of marbles did he have?

Answer: _____

21. What is the largest 3-digit number that is less than 860?

Answer: _____

5. Age Problems

5.1 Basic Word Problem on Age (*)

Example 1:

David was 8 years old in 2014. How old will he be in 2018?

Solution:

As given in the question:

David's age in 2014 = 8 years

We can use the following steps to answer the question:

Difference between 2018 and 2014

= 2018 − 2014

= 4 years

To find David's age in 2018, add 4 years to his age in 2014.

(David's age in 2018)

= (David's age in 2014) + 4

= 8 + 4

= 12 years

So David will be 12 years old in 2018.

Example 2:

Jay is currently 20 years old. How old was he 5 years ago?

Solution:

As given in the question:

Current age of Jay = 20 years

We can use the following steps to answer the question:

To find Jay's age 5 years ago, subtract 5 years from his current age.

(Jay's age 5 years ago)

= (current age of Jay) − 5

= 20 − 5

= 15 years

So 5 years ago, Jay was 15 years old.

Write the answer.

1. Bob is currently 16 years old. How old was he 6 years ago?

Answer: _____ _____
unit

2. George will be 29 years old in 4 years. How old is he now?

Answer: _____ _____
unit

Write the answer.

3. Allen is currently 15 years old. How old will he be in 7 years?

Answer: _____ _____
unit

4. Bill was 10 years old in 2010. How old will he be in 2026?

Answer: _____ _____
unit

5. Juhi is currently 35 years old. How old will she be in 11 years?

Answer: _____ _____
unit

6. Kavya was 20 years old in 2014. How old was she in 2009?

Answer: _____ _____
unit

7. Nathan will be 13 years old in 2017. How old will he be in 2024?

Answer: _____ _____
unit

8. Jack is currently 23 years old. How old was he 3 years ago?

Answer: _____ _____
unit

9. Lisa will be 17 years old in 3 years. How old is she now?

Answer: _____ _____
unit

10. Sam will be 11 years old in 2023. How old was he in 2014?

Answer: _____ _____
unit

11. Rob was 21 years old in 2009. How old will he be in 2012?

Answer: _____ _____
unit

12. Julie was 25 years old 5 years ago. How old is she now?

Answer: _____ _____
unit

5.2 Sum of and Difference between Ages (**)

Example 1:

The difference between Aditya's and Bijay's ages is 4. What will be the difference in their ages 2 years from now?

Solution:

The following information is given:
Difference between Aditya's and Bijay's current ages = 4 years

We can find the difference in their ages at a different time using the following steps:

Find the difference between their ages 2 years from now.

(Aditya's age 2 years from now)
= (Aditya's current age) + 2

(Bijay's age 2 years from now)
= (Bijay's current age) + 2

(difference between their ages in 2 years)
= (Aditya's age 2 years from now)
    − (Bijay's age 2 years from now)
= (Aditya's current age) + 2
    − (Bijay's current age) − 2
= (Aditya's current age)
    − (Bijay's current age)
= (difference between their current
                                    ages)
= 4 years

So the difference in their ages 2 years from now will be 4 years.

Note: The difference between two people's ages always remains the same. This difference is the age that the older person was when the younger person was born.

Example 2:

The sum of Lucy's and Kiran's ages is 19. What was the sum of their ages 4 years ago?

Solution:

As given in the question:

Sum of Lucy's and Kiran's ages = 19 years

We can use the following steps to answer the questions:

• Find ages 4 years ago

(Lucy's age 4 years ago)
            = (Lucy's current age) − 4
(Kiran's age 4 years ago)
            = (Kiran's current age) − 4

• Find sum of their ages 4 years ago

(sum of their ages 4 years ago)
    = (Lucy's age 4 years ago)
        + (Kiran's age 4 years ago)
    = (Lucy's current age) − 4
        + (Kiran's current age) − 4
    = (Lucy's current age)
        + (Kiran's current age) − 4 − 4
    = (Sum of Lucy's and Kiran's ages) − 8
    = 19 − 8 = 11 years

So the sum of their ages 4 years ago was 11 years.

Write the answer.

1. The difference between Jatin's and John's ages is 9. What will be the difference in their ages 4 years from now?

    Answer: _____ _____
    unit

2. The sum of Maya's and Jenny's ages is 23. What was the sum of their ages 3 years ago?

    Answer: _____ _____
    unit

3. The sum of Kunal's and Nitish's ages is 12. What will be the sum of their ages 2 years from now?

    Answer: _____ _____
    unit

4. Ayan's grandfather's age is 80. Ayan's current age is one-fourth of his grandfather's age. What is Ayan's current age?

    Answer: _____ _____
    unit

5. The difference between Neeraj's and Bob's ages is 6. What was the difference in their ages 6 years ago?

    Answer: _____ _____
    unit

6. Robert's mother's age is 36. Robert's age is half of his mother's age. What is Robert's current age?

    Answer: _____ _____
    unit

7. The sum of Jiten's and Sam's ages is 17. What was the sum of their ages 5 years ago?

    Answer: _____ _____
    unit

8. David's sister's age is 25. David's age is the same as his sister's age. What is David's current age?

    Answer: _____ _____
    unit

9. The difference between Jack's and Allen's ages is 12. What will be the difference in their ages 7 years from now?

    Answer: _____ _____
    unit

10. The sum of Rahi's and Neha's ages is 35. What will be the sum of their ages 8 years from now?

    Answer: _____ _____
    unit

5.3 Solving Age Problems at One Time—1 (**)

Example 1:

Kamal is presently 5 years old. His brother is 5 times as old as Kamal. What is his brother's age?

Solution:

As given in the question:

Current age of Kamal = 5 years
Current age of his brother

= 5 times the current age of Kamal

= 5 × 5

= 25 years

So Kamal's brother's age is 25 years.

Example 2:

John's age is half of his father's age. His father's age is 56. What is John's age?

Solution:

As given in the question:

John's father's age = 56 years
John's age = half of his father's age

$$= \frac{1}{2} \times 56$$

= 28 years

So John's age is 28 years.

Write the answer.

1. Kunal is 20 years old. He is presently 2 times as old as Allen. How old is Allen?

Answer: ____ _____
        unit

2. Jack was 18 years old in June of 2015. He was one-third of his brother's age. How old was his brother in June of 2015?

Answer: ____ _____
        unit

3. Sofia's age is one-fourth of her sister's age. Her sister's age is 12. What is Sofia's age?

Answer: ____ _____
        unit

4. Akhil's current age is two-thirds of his brother's age. His brother's age is 18. What will be Akhil's age in 9 years?

Answer: ____ _____
        unit

5. Andy's age is half of his father's age. His father's age is 48. What is Andy's age?

Answer: ____ _____
        unit

6. Arushi's age is one-third of her cousin's age. Her cousin's age is 18. What is Arushi's age?

Answer: ____ _____
        unit

Write the answer.

7. Kavya was 26 years old in May of 2016. She was half of her brother's age. How old was her brother in May of 2016?

    Answer: _____ _____
    unit

8. Jay's current age is half of his uncle's age. His uncle's age is 36. What will be Jay's age 4 years from now?

    Answer: _____ _____
    unit

9. Kapil is 27 years old. He is presently 3 times as old as Jay. How old is Jay?

    Answer: _____ _____
    unit

10. Bijay's current age is two-fifths of his father's age. His father's age is 55. What was Bijay's age 2 years ago?

    Answer: _____ _____
    unit

11. Lucy's age is half of her brother's age. Her brother's age is 28. How old is Lucy?

    Answer: _____ _____
    unit

12. Andrew is 16 years old. He is presently 4 times as old as David. How old is David?

    Answer: _____ _____
    unit

13. Julie is 12 years old. Jyoti is presently 2 times as old as Julie. How old is Jyoti?

    Answer: _____ _____
    unit

14. Angela will be 13 years old in 2018. She will be one-fourth of her mother's age. How old will her mother be in 2018?

    Answer: _____ _____
    unit

5.4 Solving Age Problems at One Time—2 (**)

Example 1:

Maria's age is half of her aunt's age. Her aunt's age is 32. Find the difference between their ages.

Solution:

As given in the question:

Maria's aunt's age = 32 years
Maria's age = half of her aunt's age

$$= \frac{1}{2} \times 32$$

$$= 16 \text{ years}$$

Difference between their ages
= Maria's aunt's age – Maria's age
= 32 – 16
= 16 years

So the difference between their ages is 16 years.

Example 2:

Nitish is presently 12 years old. His sister is 3 times as old as Nitish. What is the sum of their ages?

Solution:

As given in the question:

Current age of Nitish = 12 years
Current age of his sister

= 3 times Nitish's age

= 3 × 12

= 36 years

Sum of their ages = Nitish's age
+ His sister's age
= 12 + 36
= 48 years

So the sum of their ages is 48 years.

Write the answer.

1. Ana's age is one-fifth of her grandmother's age. Her grandmother's age is 85. Find the difference between their ages.

   Answer: _____ _____
   unit

2. Bijay's current age is two-fifths of his brother's age. His brother's age is 15. What will be Bijay's age 5 years from now?

   Answer: _____ _____
   unit

3. Disha is presently 22 years old. Her sister is 4 times as old as Disha. What is the sum of their ages?

   Answer: _____ _____
   unit

4. Jay is 16 years old. Jay is presently 2 times as old as Bob. How old is Bob?

   Answer: _____ _____
   unit

Write the answer.

5. Jack's current age is one-fifth of his grandfather's age. His grandfather's age is 75. What will be Jack's age 3 years from now?

Answer: ____ _____
unit

6. John is 32 years old. John is presently 3 times as old as Alka. How old is Alka?

Answer: ____ _____
unit

7. Olivia's age is half of her sister's age. Her sister's age is 8. Find the sum of their ages.

Answer: ____ _____
unit

8. Adriana is 27 years old. Adriana is presently 3 times as old as Anisha. How old is Anisha?

Answer: ____ _____
unit

9. Lucy is 17 years old. Disha's age is same as Lucy's age. How old is Disha?

Answer: ____ _____
unit

10. Amar's age is one-fourth of his uncle's age. His uncle's age is 60. Find the difference between their ages.

Answer: ____ _____
unit

11. Rob is presently 15 years old. His brother is 2 times as old as Rob. What is the sum of their ages?

Answer: ____ _____
unit

12. Allen is presently 26 years old. His friend's age is same as Allen's age. What is the sum of their ages?

Answer: ____ _____
unit

5.5 Review of Chapter 5 (***)

Write the answer.

1. Kunal is currently 23 years old. How old was he 5 years ago?

    Answer: _____ _____
    unit

2. Jyoti is 39 years old. Jyoti is presently two times as old as Alice. How old is Alice?

    Answer: _____ _____
    unit

3. Lucy's age is one-fourth of her mother's age. Her mother's age is 44. Find the difference between her mother's age and Lucy's age.

    Answer: _____ _____
    unit

4. Rob will be 23 years old in 5 years. How old is he now?

    Answer: _____ _____
    unit

5. Pamela's father's age is 36. Pamela's age is one-fourth of her father's age. What is Pamela's current age?

    Answer: _____ _____
    unit

6. Rose will be 21 years old in 4 years. How old is she now?

    Answer: _____ _____
    unit

7. Disha is 28 years old. Disha is presently 2 times as old as Julie. How old is Julie?

    Answer: _____ _____
    unit

8. Bill is presently 12 years old. His brother is 2 times as old as Bill. What is his brother's age?

    Answer: _____ _____
    unit

Write the answer.

9. Arnav's current age is two-fourths of his sister's age. His sister's age is 30. What will be Arnav's age 9 years from now?

Answer: _____ _____
unit

10. Ruhi will be 20 years old in 6 years. How old is she now?

Answer: _____ _____
unit

11 Anisha was 32 years old in 2015. How old will she be in 2018?

Answer: _____ _____
unit

12. Jack's age is one-fifth of his aunt's age. His aunt's age is 45. Find the difference between his aunt's age and Jack's age.

Answer: _____ _____
unit

13. Jenney is presently 29 years old. His mother is 2 times as old as Jenney. What is the sum of their ages?

Answer: _____ _____
unit

14. Arushi is 24 years old. Arushi is presently 4 times as old as Andy. How old is Andy?

Answer: _____ _____
unit

15. The sum of Maya's and Nisha's current ages is 33. What was the sum of their ages 4 years ago?

Answer: _____ _____
unit

16. Tiffany's sister's age is 42. Tiffany's age is one-third of her sister's age. What is Tiffany's current age?

Answer: _____ _____
unit

6. Time and Distance Problems

6.1 Measuring Time (*)

Example 1:

I am a clock with two hands. My hour hand is between 8 and 9. My minute hand is at 7. If somebody is looking at me in the morning, what time will I show?

(a) 9:35 a.m.
(b) 7:40 a.m.
(c) 8:35 a.m.
(d) 8:35 p.m.

Solution:

Remember the following facts while reading a clock:

- An hour hand moves from one number to the next in 1 hour. So at the half hour, the hour hand points to the middle between two numbers.

- A minute hand moves a full circle in an hour (60 minutes). So it moves from one number to the next in 5 minutes.

The hour hand is between 8 and 9, so the hour is 8.

The minute hand is at 7. Start from 12 and skip count by 5 minutes. The minute hand moves from 12 to 7 in 35 minutes. So the minute is 35.

If somebody is looking at the clock in the morning, then the clock will read as a.m.

So it will read 8:35 a.m., and the answer is (c).

Example 2:

I am a clock with three hands. My hour hand is between 10 and 11. My minute hand is between 35 and 36. If my second hand is at 18 and it is already dark outside, what time do I show?

(a) 10:36:18 p.m.
(b) 10:35:18 p.m.
(c) 10:35:18 a.m.
(d) 10:36:18 a.m.

Solution:

Remember the following facts while reading a clock:

- An hour hand moves from one number to the next in 1 hour. So at the half hour, the hour hand points to the middle between two numbers.

- 12:00 in the day is called *noon* and 12:00 in the night is called *midnight*.

- Times from midnight to noon (through the morning) are in the *a.m.,* and times from noon to midnight (through the evening) are in the *p.m.*

The hour hand is between 10 and 11, so the hour is 10.

The minute hand is between 35 and 36, so the minute is 35.

The second hand is at 18, so the second is 18.

If it is dark outside, then it is evening time and will read as p.m.

So the clock will show 10:35:18 p.m., and the answer is (b).

Write or choose the letter of the answer.

1. If it is 11:10 a.m. now, what will the time be after 3 hours?

    Answer: _____ _____
    unit

2. I am a clock with three hands. My hour hand is between 6 and 7. My minute hand is between 20 and 21. If my second hand is at 54 and it is sunny outside, what time do I show?

    (a) 7:21:54 a.m.
    (b) 6:20:54 a.m.
    (c) 6:21:54 a.m.
    (d) 7:20:54 a.m.

    Answer: _____

3. I am a clock with two hands. My hour hand is between 5 and 6. My minute hand is at 2. If somebody is looking at me in the morning, what time will I show?

    (a) 5:10 p.m.
    (b) 6:10 a.m.
    (c) 5:10 a.m.
    (d) 6:10 p.m.

    Answer: _____

4. If it is 10:05 a.m. now, what was the time 5 hours back?

    Answer: _____ _____
    unit

5. I am a clock with two hands. My hour hand is between 10 and 11. My minute hand is at 10. If somebody is looking at me in the evening, what time will I show?

    (a) 10:50 p.m.
    (b) 11:50 a.m.
    (c) 11:50 p.m.
    (d) 10:50 a.m.

    Answer: _____

6. If it is 8:55 a.m. now, what was the time 9 hours back?

    Answer: _____ _____
    unit

7. I am a clock with three hands. My hour hand is between 11 and 12. My minute hand is between 14 and 15. If my second hand is at 37 and it is dark outside, what time do I show?

    (a) 11:14:37 p.m.
    (b) 12:14:37 a.m.
    (c) 11:15:37 p.m.
    (d) 12:15:37 a.m.

    Answer: _____

8. If it is 8:09 a.m. now, what will the time be after 3 hours?

    Answer: _____ _____
    unit

6.2 Problems on Elapsed Time—1 (*)

Example 1:

A magic show started at 11:10 a.m. and ended at 11:55 a.m.

How long was the show in minutes?

Solution:

The show started at 11:10 a.m. and ended at 11:55 a.m.

So start at 11:10 a.m. and count the number of minutes until we reach 11:55 a.m.

First count by 5 minutes and then count by 1 minute.

There are 45 minutes between the start and end times.

So the show was 45 minutes.

Example 2:

Nancy returned from a 4-day summer event that ended on August 19. On which date did the event start?

Solution:

Nancy returned from a 4-day summer event. That means the event took place for 4 days.

The event ended on August 19.

So the event must be started 3 days before August 19.

The starting date of the event is = 19 − 3 = 16

So the event started on August 16.

Example 3:

If February 10 is on a Sunday, which day will it be on February 16?

Solution:

Method 1:

Write all the days in a week and date.

Days in a Week	Date
Sunday	Feb 10
Monday	Feb 11
Tuesday	Feb 12
Wednesday	Feb 13
Thursday	Feb 14
Friday	Feb 15
Saturday	Feb 16

February 16 will come after February 10.

If February 10 is on a Sunday, count up and check; February 16 will be on a Saturday.

So it will be Saturday on February 16.

Method 2:

February 10 is on Sunday.

February 16 will come after 6 days from February 10.

So we have to count 6 days from Sunday (February 10) to reach February 16.

Start with February 10: Sunday

February 11: Monday

-

-

-

February 16: Saturday

So it will be Saturday on February 16.

Write the answer.

1. A cricket match started at 4:00 p.m. and ended at 7:00 p.m.

    How long was the match in hours?

    Answer: ____ _____
    unit

2. If December 24 is on a Wednesday, which day will it be on December 27?

    Answer: _____

3. Paul went to a circus at 7:30 p.m. and returned at 9:10 p.m.

    How long did he stay at the circus?

    Answer: ____ _____
    unit

4. A dance competition started at 3:30 p.m. and ended at 4:25 p.m.

    How long was the competition in minutes?

    Answer: ____ _____
    unit

5. Maya went to a shopping mall at 12:22 p.m. and returned at 2:25 p.m.

    How long did she stay at the shopping mall?

    Answer: ____ _____
    unit

6. Gabriel returned from a 10-day cultural event that ended on July 16.

    On which date did the event start?

    Answer: ____ _____
    unit

7. A birthday celebration started at 6:12 p.m. and ended at 10:12 p.m.

    How long was the celebration in hours?

    Answer: ____ _____
    unit

8. If April 12 is on a Tuesday, which day will it be on April 21?

    Answer: _____

6.3 Problems on Elapsed Time—2 (*)

Example 1:

Jack went on a tour with his friends on October 3 and returned on October 14. How many days was Jack gone for the tour?

Solution:

Jack went on a tour on October 3 and returned on October 14.

Number of days Jack was gone for the tour
= the day he returned – the day he went
= October 14 – October 3
= 14 – 3 = 11 days

So Jack was gone 11 days for the tour.

Example 2:

Rob is visiting his cousin in a different city and wants to stay for 14 days at his cousin's place. If he starts on May 25, what day will he return?

Solution:

The following information is given:

Rob visits his cousin on May 25.
Number of days he wants to stay = 14

So he will return after 14 days from May 25.

There are 31 days in May.

Rob will stay 31 – 25 = 6 days in May and another 14 – 6 = 8 days in the next month, June.

So Rob will return on June 8.

Example 3:

An opening ceremony of a hospital started at 11:00 a.m. and ended at 2:40 p.m. How long was the ceremony?

(a) 3 hours 40 minutes
(b) 1 hour 40 minutes
(c) 3 hours 20 minutes
(d) 1 hour 20 minutes

Solution:

The ceremony started at 11:00 a.m. and ended at 2:40 p.m.

So start at 11:00 a.m. and count the number of hours until we reach 2:40 p.m.

First count by hours, and then count by minutes.

There are 3 hours and 40 minutes between the start and end times.

So the ceremony lasted 3 hours and 40 minutes.

Write or choose the letter of the answer.

1. An annual function of a school started at 10:00 a.m. and ended at 6:25 p.m.

   How long was the function?

   (a) 6 hours 25 minutes
   (b) 8 hours 25 minutes
   (c) 10 hours 25 minutes
   (d) 4 hours 25 minutes

   Answer: _____

2. Anuj is visiting his friend and wants to stay for 3 days at his friend's place. If he starts on January 19, what day will he return?

   Answer: ____ _____
   unit

3. A movie in a movie theater started at 6:00 p.m. and ended at 8:10 p.m.

   How long was the movie?

   (a) 6 hours 00 minutes
   (b) 14 hours 10 minutes
   (c) 2 hours 10 minutes
   (d) 8 hours 10 minutes

   Answer: _____

4. Mr. Harper is visiting his guest house and stays there for 5 days. If he starts on June 15, what day will he return?

   Answer: _____

5. Jay went on a vacation on July 14 and returned on July 23.

   How many days was Jay gone for the vacation?

   Answer: ____ _____
   unit

6. A baseball game started at 5:00 p.m. and ended at 8:30 p.m.

   How long was the game?

   (a) 3 hours 30 minutes
   (b) 13 hours 30 minutes
   (c) 5 hours 30 minutes
   (d) 8 hours 30 minutes

   Answer: _____

7. Peter is visiting his uncle in a different town and wants to stay for 20 days at his uncle's place. If he starts on November 7, what day will he return?

   Answer: ____ _____
   unit

8. Julie went on a holiday tour with her parents on May 15 and returned on May 27.

   How many days was Julie gone for the holiday tour?

   Answer: ____ _____
   unit

6.4 Find Speed Given Distance and Time (**)

Example 1:

Robert drove 90 miles in 2 hours. What was his speed in miles per hour?

Solution:

The following information is given:

Distance = 90 miles

Time = 2 hours

Find the speed in miles per hour.

$$\text{Speed} = \frac{\text{distance}}{\text{time}}$$

$$= \frac{90 \text{ miles}}{2 \text{ hours}}$$

$$= \frac{90}{2} \frac{\text{miles}}{\text{hour}}$$

$$= \frac{45}{1} \frac{\text{miles}}{\text{hour}}$$

$$= 45 \text{ miles per hour}$$

So Robert's speed was 45 miles per hour.

Example 2:

Samir's family had to drive 110 miles to go for a picnic. They started from their home at 8:00 a.m. and reached the picnic place at 10:00 a.m. What was their driving speed in miles per hour?

Solution:

The following information is given:

Distance = 110 miles

Start time = 8:00 a.m.

End time = 10:00 a.m.

Start at 8:00 a.m. and count the number of hours until we reach 10:00 a.m. There are 2 hours between the start and end times.

Time = 2 hours

Find the speed in miles per hour.

$$\text{Speed} = \frac{\text{distance}}{\text{time}}$$

$$= \frac{110 \text{ miles}}{2 \text{ hours}}$$

$$= \frac{110}{2} \frac{\text{miles}}{\text{hour}}$$

$$= \frac{55}{1} \frac{\text{miles}}{\text{hour}}$$

$$= 55 \text{ miles per hour}$$

So their driving speed was 55 miles per hour.

Write the answer.

1. Anil ran 24 miles in 3 hours. What was his speed in miles per hour?

   Answer: _____ _____
                       unit

2. Mr. Rao's family had to travel 135 kilometers to go to a marriage party. They started from their home at 5:00 p.m. and reached the party at 8:00 p.m. What was their speed in kilometers per hour?

   Answer: _____ _____
                       unit

3. Nancy's college is 600 meters from her home. If she took 15 minutes to walk from her home to college, what was her speed in meters per minute?

   Answer: _____ _____
                       unit

4. A butterfly flew 20 miles in 2 hours. What was its speed in miles per hour?

   Answer: _____ _____
                       unit

5. A biker had to ride 160 miles to reach his destination. He started from his home at 10:00 a.m. and reached the destination at 2:00 p.m. What was his speed in miles per hour?

   Answer: _____ _____
                       unit

6. Luke's school is 300 meters from his home. If he took 10 minutes to walk from his home to school, what was his speed in meters per minute?

   Answer: _____ _____
                       unit

7. Jessica ran an 1,800-meter marathon in 15 minutes. What was her speed in meters per minute?

   Answer: _____ _____
                       unit

8. Rob had to travel 90 miles to go to his friend's house by bike. He started from his home at 6:00 a.m. and reached his friend's house at 9:00 a.m. What was his speed in miles per hour?

   Answer: _____ _____
                       unit

9. Mrs. Cook's town park is 210 meters from her home. If she took 6 minutes to walk from her home to park, what was her speed in meters per minute?

   Answer: _____ _____
                       unit

10. A bullet was fired at 960 meters in 3 seconds. What was its speed in meters per second?

    Answer: _____ _____
                        unit

6.5 Find Distance Given Speed and Time (**)

Example 1:

A boat can travel 8 kilometers in 1 hour. How much distance will it cover in 3 hours?

Solution:

The following information is given:

The boat travels 8 kilometers in 1 hour

Speed = 8 kilometers per hour

Time = 3 hours

You can find the distance using the following formula:

Distance = (speed) × (time)

= 8 × 3

= 24 kilometers

So the boat will cover a distance of 24 kilometers in 3 hours.

Example 2:

Albert was driving at a speed of 50 miles per hour. If he took half an hour to go from his home to office, how far is his office from his home?

Solution:

The following information is given:

Speed = 50 miles per hour

Time taken by Albert = half an hour

$\text{Time} = \dfrac{1}{2}$ hour

You can find the distance using the following formula:

Distance = (speed) × (time)

$= 50 \times \dfrac{1}{2}$

$= \dfrac{50}{2} = 25$ kilometers

So Albert's office is 25 kilometers away from his home.

Write the answer.

1. Roni's family and relatives took a bus to go for a tour. They left home at 6:00 a.m. and reached the place at 10:00 a.m. If the bus was traveling at a speed of 60 kilometers per hour, how far is the place from their house?

2. Soham was walking at a speed of 75 meters per minute. If he took 10 minutes to go from his home to the gym, how far is the gym from his home?

Answer: _____ _____
unit

Answer: _____ _____
unit

Write the answer.

3. David was driving at a speed of 48 kilometers per hour. If he took one-and-a-half hours to go from his office to the museum, how far is the museum from his office?

    Answer: _____ _____

    unit

4. Roshni's friends took a car to go to a dance camp. They left school at 12:00 p.m. and reach at the camp at 2:00 p.m. If the car was traveling at a speed of 60 miles per hour, how far is the camp from their school?

    Answer: _____ _____

    unit

5. A dolphin can swim 9 miles in 1 hour. How much distance will it cover in 5 hours?

    Answer: _____ _____

    unit

6. Daniel was running at a speed of 12 kilometers per hour in a race. If he took half an hour to complete his race, how far is the race?

    Answer: _____ _____

    unit

7. Tania went to her dance class by walking. She left home at 4:00 p.m. and reached the class at 4:10 p.m. If she was walking at a speed of 60 meters per minute, how far is the dance class from her home?

    Answer: _____ _____

    unit

8. Franc was traveling at a speed of 40 miles per hour. If he took half an hour to go from his home to college, how far is his college from his home?

    Answer: _____ _____

    unit

9. Nikita's school class took a bus to go for an excursion. They left school at 8:00 a.m. and reached the place at 11:00 a.m. If the bus was traveling at a speed of 30 miles per hour, how far is the distance they traveled from their school?

    Answer: _____ _____

    unit

10. A horse can run 45 kilometers in 1 hour. How much distance will it cover in 5 hours?

    Answer: _____ _____

    unit

6.6 Find Time Given Speed and Distance (**)

Example 1:

Rohan runs 200 meters in 1 minute. If he has to run for 3 kilometers and 400 meters, how much time will he take?

Solution:

The following information is given:

We know that 1 kilometer = 1,000 meters

3 kilometers = 3 × 1,000 meters
= 3,000 meters

Distance = 3 kilometers and 400 meters
= 3,000 meters + 400 meters
= 3,400 meters

Speed = 200 meters per minute

You can find the time using the following formula:

Distance = (Speed) × (Time)

Time = (Distance) ÷ (Speed)
= 3,400 ÷ 200
= 17 minutes

So Rohan will take 17 minutes to run for 3 kilometers and 400 meters.

Example 2:

Nikhil's office is 80 miles from his home. If he drives at 40 miles per hour and starts from his home at 7:00 a.m., what time will he arrive at his office?

Solution:

The following information is given:

Distance = 80 miles
Speed = 40 miles per hour
Start time = 7 a.m.

You can find the time using the following formula:

Distance = (Speed) × (Time)

Time = (Distance) ÷ (Speed)
= 80 ÷ 40
= 2 hours

There are 2 hours between the start and end times.

Start at 7:00 a.m. and count up to 2 hours and we reach 9:00 a.m.

So Nikhil will arrive at his office at 9:00 a.m.

Write the answer.

1. A train travels 40 miles in 1 hour. How much time will it take to travel 240 miles?

Answer: _____ _____
unit

2. Joy's shop is 800 meters from his home. If he walks at 40 meters per minute and starts from his home at 8:00 a.m., what time will he arrive at his shop?

Answer: _____ _____
unit

Write the answer.

3. Manoj runs 150 meters in 1 minute. If he has to run for 1 kilometer and 350 meters, how much time will he take?

Answer: _____ _____
unit

4. Kyle's restaurant is 40 miles from his home. If he drives at 20 miles per hour and starts from his home at 9:00 a.m., what time will he reach his restaurant?

Answer: _____ _____
unit

5. A truck travels 45 kilometers in 1 hour. How much time will it take to travel 290 kilometers?

Answer: _____ _____
unit

6. A car travels 800 meters in 1 minute. If it has to travel for 8 kilometers and 800 meters, how much time will it take?

Answer: _____ _____
unit

7. Andy drives 35 kilometers in 1 hour. How much time will he take to travel 280 kilometers?

Answer: _____ _____
unit

8. Sarika's art class is 500 meters from her home. If she walks at 50 meters per minute and starts from her home at 7:30 a.m., what time will she reach her class?

Answer: _____ _____
unit

9. A car travels 500 meters in 1 minute. If it has to travel for 2 kilometers and 500 meters, how much time will it take?

Answer: _____ _____
unit

10. An airplane covers 900 kilometers in 1 hour. How much time will it take to cover 3,600 kilometers?

Answer: _____ _____
unit

11. Neha's home is 60 miles from her club. If she drives at 20 miles per hour and starts from her club at 4:00 p.m., what time will she reach her home?

Answer: _____ _____
unit

12. Mark runs 3 meters in 1 second. How much time will he take to travel 150 meters?

Answer: _____ _____
unit

6.7 Review of Chapter 6 (**)

Write or choose the letter of the answer.

1. I am a clock with three hands. My hour hand is between 3 and 4. My minute hand is between 18 and 19. If my second hand is at 15 and it is hot outside, what time do I show?
   - (a) 4:18:15 p.m.
   - (b) 3:18:15 p.m.
   - (c) 4:19:15 a.m.
   - (d) 3:19:15 a.m.

   Answer: _____

2. If August 6 is on a Monday, which day will it be on August 20?

   Answer: _____

3. Niki's parlor is 700 meters from her home. If she walks at 45 meters per minute and starts from her home at 9:00 a.m., what time will she reach her parlor?

   Answer: ____ _____
   unit

4. Stephen drove 105 miles in 3 hours. What was his speed in miles per hour?

   Answer: ____ _____
   unit

5. If it is 2:45 p.m. now, what was the time 4 hours ago?

   Answer: ____ _____
   unit

6. An opening ceremony of a shopping mall started at 10:30 a.m. and ended at 1:30 p.m. How long was the ceremony?
   - (a) 2 hours
   - (b) 1 hour
   - (c) 4 hours
   - (d) 3 hours

   Answer: _____

7. Jay's family had to drive 80 miles to go to a museum. They started from their home at 7:00 a.m. and reached the museum at 9:00 a.m. What was their driving speed in miles per hour?

   Answer: ____ _____
   unit

8. A tiger runs 45 kilometers in 1 hour. If it has to run for 225 kilometers, how much time will it take?

   Answer: ____ _____
   unit

Write or choose the letter of the answer.

9. If April 12 is on a Wednesday, which day will it be on April 18?

    Answer: _____

10. Linda runs 150 meters in 1 minute. If she has to run for 3 kilometers and 300 meters, how much time will she take?

    Answer: _____ _____
    <small>unit</small>

11. If it is 11:00 a.m. now, what was the time 3 hours back?

    Answer: _____ _____
    <small>unit</small>

12. Nancy's friends had to drive 220 kilometers to go for a picnic. They started from their home at 6:00 a.m. and reached at the picnic place at 10:00 a.m. What was their driving speed in kilometers per hour?

    Answer: _____ _____
    <small>unit</small>

13. A movie show started at 12:40 p.m. and ended at 3:00 p.m.

    How long was the show in minutes?

    Answer: _____ _____
    <small>unit</small>

14. Nikhil's college is 100 miles from his home. If he drives at 50 miles per hour and starts from his home at 6:30 a.m., what time will he reach his college?

    Answer: _____ _____
    <small>unit</small>

15. A welcome ceremony of a college started at 9:00 a.m. and ended at 3:30 p.m. How long was the ceremony?

    (a) 6 hours 30 minutes
    (b) 3 hours 30 minutes
    (c) 5 hours 20 minutes
    (d) 6 hours 15 minutes

    Answer: _____

16. A car traveled 160 miles in 5 hours. What was its speed in miles per hour?

    Answer: _____ _____
    <small>unit</small>

17. I am a clock with two hands. My hour hand is between 6 and 7. My minute hand is at 9. If somebody is looking at me in the evening, what time will I show?

    (a) 6:35 a.m.
    (b) 7:45 a.m.
    (c) 6:45 p.m.
    (d) 7:30 p.m.

    Answer: _____

7. Money Problems

7.1 Shopping Problem (*)

Example 1:

Ravi bought 5 shirts for $15.00 each and 3 pairs of jeans for $22.50 each from a store. How much money did he pay in total?

Solution:

The following information is given:

Cost of 1 shirt = $15.00
Number of shirts bought = 5
Cost of 1 pair of jeans = $22.50
Number of jeans bought = 3

We can find the answer as given below.

- Cost of 5 shirts

 = cost of 1 shirt × number of shirts

 = $15.00 × 5

 = $75.00

- Cost of 3 pairs of jeans

 = cost of 1 pair × number of pairs

 = $22.50 × 3

 = $67.50

Total amount paid

 = cost of 5 shirts + cost of 3 pairs of jeans

 = $75.00 + $67.50 = $142.50

So Ravi paid $142.50 in total.

Example 2:

Samir ordered 10 packs of rice noodles and 12 packs of wheat noodles and paid a total of $9.30. If the cost of each pack of rice noodles is $0.45, what is the cost of each pack of wheat noodles?

Solution:

The following information is given:

Number of rice noodle packs = 10
Number of wheat noodle packs = 12
Total amount paid = $9.30
Cost of 1 rice noodle pack = $0.45

We can use the following steps to find the answer:

Step 1: Find the total cost of rice noodles.

(total cost of rice noodles)
 = (number of rice noodle packs)
 × (cost of 1 rice noodle pack)
 = $10.00 × $0.45 = $4.50

Step 2: Find the amount paid for wheat noodle packs.

(amount paid for wheat noodle packs)
 = (total amount paid)
 − (total cost of rice noodles)
 = $9.30 − $4.50 = $4.80

Step 3: Find the cost of 1 wheat noodle pack.

(cost of 1 wheat noodle pack)
 = (amount paid for wheat noodle packs)
 ÷ (number of wheat noodle packs)
 = $4.80 ÷ 12 = $0.40

So each wheat noodle pack costs $0.40.

Write the answers.

1. Mr. Hales bought 3 caps for $2.00 each and 2 backpacks for $15.00 each from a store for his children. How much money did he pay in total?

    Answer: _____

2. Ana ordered 1 large pizza that costs $20.00. She also ordered 4 bread sticks. The price of bread sticks is $0.50 per stick. If Ana gave $30.00 to the cashier, how much money did the cashier return?

    Answer: _____

3. Adriana ordered 8 packets of cashews and 10 packets of almonds and paid a total of $200.00. If the cost of each packet of cashews is $10.00, what is the cost of each packet of almonds?

    Answer: _____

4. Julie went to watch a movie. She spent $8.50 for the movie ticket, $7.50 for dinner, and $3.75 for popcorn. If she had $2.25 left at the end, how much money did she have at the beginning?

    Answer: _____

5. Bill ordered 1 keyboard that costs $18.00. He also ordered 3 headphones. The price of each headphone set is $9.00. If Bill gave $50.00 to the cashier, how much money did the cashier return?

    Answer: _____

6. Liza ordered 4 ice-cream cones and 8 pizzas and paid a total of $42.00. If the cost of each ice-cream cone is $2.50, what is the cost of each pizza?

    Answer: _____

7. Peter went on an outing with his friends. He spent $1.50 for transportation, $8.50 for lunch, and $15.25 for clothes. If he had $4.75 left at the end, how much money did he have at the beginning?

    Answer: _____

8. Sujie bought 4 wallets for $5.00 each and 2 chairs for $18.50 each from a store. How much money did she pay in total?

    Answer: _____

7.2 Expense Planning (**)

Example 1:

Lisa wants to buy a bicycle that costs $85.00. She also needs to buy school supplies that cost $18.00, a backpack that costs $22.00, and a lunch that costs $8.00. How much money does she need in total?

Solution:

The following information is given:

Cost of a bicycle = $85.00
Cost of school supplies = $18.00
Cost of a backpack = $22.00
Cost of lunch = $8.00

We can find the answer as given below.

Total amount of money Lisa needs

= cost of a bicycle
   + cost of school supplies
   + cost of a backpack
   + cost of lunch
= $85.00 + $18.00 + $22.00 + $8.00
= $133.00

So Lisa needs $133.00 in total.

Example 2:

Simon will be going to college next year. He needs to pay $8,500.00 for tuition fee per year and $625.00 per month for room and board. He also needs an additional $1,250.00 per year to buy textbooks and supplies. If he needs to stay in college for 4 years for his engineering degree, how much total money does he need for the degree?

Solution:

The following information is given:

Tuition fee per year = $8,500.00
Room and board charges
  = $625.00 per month
  = $625.00 × 12 = $7,500.00 per year
Additional expenses per year
                  = $1,250.00
Period = 4 years

We can find the answer as given below.

First, we need to find the total expenses for 1 year.

Total expenses for 1 year = Tuition fee
              + Room and board charges
              + Additional expenses
   = $8,500.00 + $7,500.00 + $1,250.00
   = $17,250.00

Then, we need to find the total expenses for 4 years.

Total expenses for 4 years
          = Total expenses for 1 year × 4
          = $17,250.00 × 4 = $69,000.00

So Simon needs a total of $69,000.00 for the engineering degree.

Name _____

Write the answer.

1. Nancy is planning for her brother's birthday party. She prepared the following list of things she needs to buy and the cost of each item. How much total money does she need to ask her mother for the birthday celebration?

Item	Cost	Total Cost
Decorations	$30.00	
Cake	$15.00	
Paper plates	$10.00	
Goody bags for 20 kids	$3.00 per kid	
4 large pizzas	$18.00 per pizza	
5 bottles of soft drinks	$2.00 per bottle	

Answer: _____

2. Lora wants to buy a ring that costs $130.00. She also needs to buy a saree that costs $35.00 and a make-up kit that costs $15.00. How much money does she need in total?

Answer: _____

3. Vinod wants to buy a cooler that costs $150.00. He also needs to buy a laptop that costs $500.00. How much money does he need in total?

Answer: _____

4. Charl will be starting a course next month. He needs to pay $400.00 for course fee per month and $60.00 per month for room and board. He also needs an additional $50.00 per month to buy textbooks and supplies. If he needs to do the course for 6 months, how much total money does he need for the entire period of the course?

Answer: _____

5. Mia wants to buy a carpet that costs $45.00. She also needs to buy home expenses that cost $25.00, a coffee mug that costs $6.00, and sunglasses that costs $20.00. How much money does she need in total?

Answer: _____

6. Mr. Smith will be placing some workers for fieldwork for a week. He needs to pay them $50.00 per day for daily charges and $10.00 per day for lunch. He also pays an additional $5.00 per day for their personal expenses. If Mr. Smith needs to pay them for 7 days for his work, how much total money does he need to spend on them?

Answer: _____

7.3 Investment Problem (***)

Example 1:

Sam and his friend started a small business with an investment of $300.00. After 2 years they have a balance of $642.00 in their bank account. How much profit did they make in 2 years?

Solution:

The following information is given:

Amount of money invested = $300.00
Time period = 2 years
Balance after 2 years = $642.00

We can find the answer as given below.

Profit can be calculated by subtracting the investment amount from the balance after 2 years.

Profit = (balance after 2 years)
— (amount of money invested)
= $642.00 – $300.00 = $342.00

So they made a profit of $342.00 in 2 years.

Example 2:

John has $500.00 in his savings account at the bank. He took $250.00 from this account, invested it in a mutual fund, and earned a $20.00 profit in 1 year. He earned $5.00 in interest for the remaining money in the savings account. How much total money does John have after 1 year?

Solution:

The following information is given:

Amount in savings account = $500.00
Amount invested in mutual fund
= $250.00
Profit earned from mutual fund in 1 year
= $20.00
Interest earned for the remaining money in savings account = $5.00

Total balance in mutual fund
= Amount invested + profit
= $250.00 + $20.00 = $270.00

Remaining money in savings account
= Amount that was in savings account
— amount invested in mutual fund
= $500.00 – $250.00 = $250.00

Total balance in savings account
= remaining money + interest
= $250.00 + $5.00 = $255.00

Total money John received
= Total balance in mutual fund
+ Total balance in savings account
= $270.00 + $255.00 = $525.00

So John has a total of $525.00 after 1 year.

Write the answer.

1. Roshni deposited an amount of $780.00 in her savings bank account. After 5 years she had a balance of $1,000.00 in her bank account. How much profit did she have after 5 years?

Answer: _____

2. Anil has invested $500.00 in a company stock. His investment grew to $650.00 in 5 years. If he has earned same amount every year, how much was his earning per year?

Answer: _____

3. Dev has $675.00 with him. He gave $350.00 to his friend and earned a $15.00 profit in 1 year. He earned $10.00 in interest on his remaining money. How much total money does Dev have after 1 year?

Answer: _____

4. Nancy deposited an amount of $430.00 in a savings fund. After 3 years she received a balance of $580.00. How much profit did she get in 3 years?

Answer: _____

5. Randal is saving for his college expenses and has $10,000.00 for investment. He has selected a mutual fund that will give a return of $100.00 per year for each $1,000.00 investment. What will be Randal's balance at the end of 10 years?

Answer: _____

6. Steve has $1,200.00 in his savings account in a bank. He took $800.00 from this account, invested it in a mutual fund, and earned a $120.00 profit in 1 year. He earned $40.00 in interest for the remaining money in the savings account. How much total money does Steve have after 1 year?

Answer: _____

7. Mrs. Waller has invested $940.00 in a company stock. Her investment grew to $1,650.00 in 2 years. If she has earned the same amount every year, how much were her earnings per year?

Answer: _____

7.4 Pricing Problem (***)

Example 1:

The cost of a pack of ice-cream cones is $3.60. The sales tax on each pack is $0.60. If Rob buys 2 packs, how much money will he pay in total including the sales tax?

Solution:

The following information is given:

Cost of 1 ice-cream cone pack = $3.60

Sales tax on 1 pack = $0.60

Number of packets bought = 2

Total cost of 1 ice-cream cone pack

= cost of 1 ice-cream cone pack
+ sales tax on 1 pack

= $3.60 + $0.60 = $4.20

Total cost of 2 ice-cream cone packs

= total cost of 1 ice-cream cone
pack × 2

= $4.20 × 2 = $8.40

So Rob will pay $8.40 in total including the sales tax.

Example 2:

Andy bought 50 pounds of apples and paid a total of $50.00. When he sells the apples, he needs to make a profit of $0.75 for each pound and charge $0.15 per pound for sales tax. How much will he be charging customers for each pound of apples?

Solution:

The following information is given:

Weight of apples bought = 50 pounds

Amount paid for apples = $50.00

Profit Andy needs = $0.75 per pound

Sales tax = $0.15 per pound

Amount of profit per pound

= Profit Andy needs + Sales tax
= $0.75 + $0.15 = $0.90

Total amount of profit

= weight of apples bought
× amount of profit per pound

= 50 × $0.90 = $45.00

Selling price of apples

= amount paid for apples
+ total amount of profit

= $50.00 + $45.00 = $95.00

Customer's price for apples per pound

= selling price of apples
÷ weight of apples bought

= $95.00 ÷ 50 = $1.90

So Andy will be charging customers $1.90 for each pound of apples.

Write the answer.

1. Joy bought 65 pairs of shoes in his shop and paid a total of $260.00. How much did he pay for each pair of shoes?

   Answer: _____

2. The cost of 1 pizza is $6.20. The sales tax on each pizza is $0.50. If Bill buys 4 pizzas, how much money will he pay in total including the sales tax?

   Answer: _____

3. Kevin bought 60 pens and paid a total of $120.00. When he sells the pens, he needs to make a profit of $0.50 for each pen and charge $0.20 per pen for sales tax. How much will he be charging customers for each pen?

   Answer: _____

4. Suman had 250 packs of pencils in her store. She sold all of them for $500.00 and made a $0.75 profit per pack of pencils. If she collected a sales tax of $0.08 per pack, how much did she pay to buy 1 pack of pencils?

   Answer: _____

5. Sarah bought 50 bottles and paid a total of $100.00. When she sells the bottles, she needs to make a profit of $0.60 for each bottle and charge $0.20 per bottle for sales tax. How much will she be charging customers for each bottle?

   Answer: _____

6. Jack bought 35 shirts in his shop and paid a total of $280.00. How much did he pay for each shirt?

   Answer: _____

7. Elina had 60 packs of sugar in her store. She sold all of them for $180.00 and made $1.00 profit per pack of sugar. If she collected a sales tax of $0.50 per pack, how much did she pay to buy 1 pack of sugar?

   Answer: _____

8. The cost of 1 box of candies is $4.00. The sales tax on each box is $0.50. If Jia buys 2 boxes, how much money will she pay in total including the sales tax?

   Answer: _____

7.5 Profit and Loss (***)

Example 1:

Jenny bought a water purifier for $210.00 and sold it for $265.00. How much profit did she make?

Solution:

The following information is given:

Cost price of water purifier = $210.00
Selling price of the purifier = $265.00

The profit can be calculated by subtracting the cost price from the selling price.

Profit = selling price – cost price
  = $265.00 – $210.00
  = $55.00

So Jenny made a profit of $55.00.

Example 2:

Bob bought a bike and sold it with an $18.00 loss because he did not like the color of the bike. If he sold the bike for $77.00, how much did he pay to buy the bike?

Solution:

The following information is given:

Loss amount = $18.00

Selling price of the bike = $77.00

The amount paid to buy the bike is the cost price of the bike.

Cost price of the bike

= Selling price of the bike + loss amount
= $77.00 + $18.00
= $95.00

So Bob paid $95.00 to buy the bike.

Example 3:

During the month of August, a store made a $500.00 *profit* on selling bikes, a $1,300.00 *profit* on selling school supplies, a $2,500.00 *profit* on selling textbooks, and a $280.00 *loss* on selling toys. How much profit did the store make in total in August?

Solution:

The following information is given:

Profit made on selling bikes = $500.00
Profit on selling school supplies
     = $1,300.00
Profit on selling textbooks = $2,500.00
Loss on selling toys = $280.00

Profit means a positive sign and *loss* means a negative sign.

Total profit = Profit made on selling bikes
  + Profit on selling school supplies
  + Profit on selling textbooks
  = $500.00 + $1,300.00 + $2,500.00
  = $4,300.00

The store made a loss of $280.00 on selling toys.

Profit = total profit – $280.00
  = $4,300.00 – $280.00
  = $4,020.00

So the store made a total profit of $4,020.00 in August.

Write the answer.

1. Mark bought a vacuum cleaner for $180.00 and sold it for $215.00. How much profit did he make?

    Answer: _____

2. Susil has spent $500.00 to buy 100 T-shirts for a school fund-raising event. If he wants to raise $700.00 for the school, how much should he ask for each T-shirt?

    Answer: _____

3. During the year of 2015, a store made a $2,500.00 profit on selling cashews, a $2,100.00 profit on selling almonds, a $2,000.00 profit on selling walnuts, and a $780.00 loss on selling cherries. How much profit did the store make in total in 2015?

    Answer: _____

4. Nigam has spent $156.00 to buy 52 caps for a school sports event. If he wants to raise $104.00 for the school, how much should he ask for each cap?

    Answer: _____

5. Dennis bought a laptop and sold it with a $600.00 loss because he did not like the model of the laptop. If he sold the laptop for $2,460.00, how much did he pay to buy the laptop?

    Answer: _____

6. Miley bought a phone for $190.00 and sold it for $165.00. How much loss did she experience?

    Answer: _____

7. Anil bought a backpack and sold it with a $9.00 loss because he did not like the size of the backpack. If he sold the backpack for $43.00, how much did he pay to buy the backpack?

    Answer: _____

8. During the month of May, a shopping mall made a $340.00 profit on selling dresses, a $780.00 profit on selling electronics items, and a $310.00 loss on selling footwears. How much profit did the mall make in total in May?

    Answer: _____

7.6 Review of Chapter 7 (**)

Write the answer.

1. Nikhil bought a headset and sold it with $7.50 loss because he did not like the headset. If he sold the headset for $22.00, how much did he pay to buy the headset?

   Answer: _____

2. The cost of a candy packet is $4.20. The sales tax on each pack is $0.30. If Kunal buys 2 packets, how much money will he give in total including the sales tax?

   Answer: _____

3. Peter ordered 12 sandwiches and 15 burgers and paid a total of $24.60. If the cost of each sandwich is $0.80, what is the cost of each burger?

   Answer: _____

4. Jimmy bought a digital camera for $140.00 and sold it for $172.00. How much profit did he make?

   Answer: _____

5. Nancy will be going to college next year. She needs to pay $2,500.00 for tuition per year and $650.00 per month for room and board. She also needs an additional $300.00 per year to buy textbooks and supplies. If she needs to stay in college for 2 years for her MBA course, how much total money does she need for the course?

   Answer: _____

6. Lima wants to buy a watch that costs $60.00. She also needs to buy a wallet that costs $11.00, a uniform that costs $32.00, and some snacks that cost $4.50. How much money does she need in total?

   Answer: _____

7. During the month of June, a jewellery shop made a $3,200.00 profit on selling gold, a $890.00 profit on selling silver, a $1,800.00 profit on selling platinum, and a $1,300.00 loss on selling diamonds. How much profit did the shop make in total in June?

   Answer: _____

Name _____

Write the answer.

8. Sunny and his friend started a small business with an investment of $500.00. After 3 years they have a balance of $740.00 in their bank account. How much profit did they make in 3 years?

Answer: _____

9. The cost of a washing machine is $210.00. The sales tax on each machine is $3.50. If Nil orders 3 machines, how much money will he give in total including the sales tax?

Answer: _____

10. Lisa wants to buy a pair of shoes that costs $40.00. She also needs to buy school supplies that cost $15.00, a water bottle that costs $7.50, and a bracelet that costs $5.00. How much money does she need in total?

Answer: _____

11. Shaun bought 3 coffee mugs for $2.50 each and 4 jugs for $4.50 each from a store. How much money did he pay in total?

Answer: _____

12. Maria bought sunglasses for $45.00 and sold them for $39.50. How much loss did she experience?

Answer: _____

13. Kile spent $480.00 to buy 40 stools for the lab in a college. If he wants to raise $520.00 for the college, how much should he ask for each stool?

Answer: _____

14. Jack bought 20 pairs of spectacles and paid a total of $400.00. When he sells the spectacles, he needs to make a profit of $1.50 for each pair of spectacles and charge $1.25 per pair of spectacles for sales tax. How much will he be charging customers for each pair?

Answer: _____

15. Emily bought 16 red bottles and 20 blue bottles and paid a total of $50.00. If the cost of each red bottle is $1.25, what is the cost of each blue bottle?

Answer: _____

8. Work Problems

8.1 Concepts of Work and Time—1 (*)

Example 1:

6 people can finish a job in 5 days. How many people will finish the job in 10 days?

Solution:

This problem can be solved using the following steps:

Step 1: Find the number of people required to finish the job in 1 day.

Number of people to finish a job in 5 days
= 6 people

Number of people to finish the job in 1 day
= 6 × 5
= 30 people

Step 2: Find the number of people required to finish the job in 10 days.

Number of people to finish the job in 1 day
= 30
Number of people to finish the job in 10 days = 30 ÷ 10
= 3 people

So 3 people will finish the job in 10 days.

Note: For a given task, more people finish the task in less time, and fewer people finish the task in more time.

Example 2:

10 people can make 40 toys in 1 day. How many toys can 6 people make in 1 day?

Solution:

This problem can be solved using the following steps:

Step 1: Find the number of toys that can be made by 1 person in 1 day.

Number of toys that can be made by 10 people in 1 day = 40

Number of toys that can be made by 1 person in 1 day = 40 ÷ 10 = 4 toys

Step 2: Find the number of toys that can be made by 6 people in 1 day.

Number of toys that can be made by 1 person in 1 day = 4
Number of toys that can be made by 6 people in 1 day = 4 × 6
= 24 toys

So 6 people can make 24 toys in 1 day.

Note: For a given time, more people make more toys, and fewer people make less toys.

Write the answer.

1. It takes 15 seconds to fill a bottle. How long will it take to fill 6 bottles?

    Answer: ____ _____
    unit

2. 4 workers can paint a wall in 3 hours. How many workers will paint the wall in 2 hours?

    Answer: ____ _____
    unit

3. 10 children drink 20 liters of water in a day. What amount of water will 1 child drink in a day?

    Answer: ____ _____
    unit

4. 8 students can make 72 paper crafts in 1 day. How many paper crafts can 12 students make in 1 day?

    Answer: ____ _____
    unit

5. 5 students complete 15 assignments in a day. How many assignments will 1 student complete in a day?

    Answer: ____ _____
    unit

6. 8 carpenters can make 32 beds in 1 month. How many beds can 5 carpenters make in 1 month?

    Answer: ____ _____
    unit

7. 5 robots can assemble a car in 6 hours. How many robots will assemble the car in 10 hours?

    Answer: ____ _____
    unit

8. A pump takes 40 minutes to fill a water tank. How long will it take to fill 3 tanks?

    Answer: ____ _____
    unit

8.2 Concepts of Work and Time—2 (*)

Example 1:

Mr. Parker can design 6 paintings in 12 days. How many days will he take to design 1 painting?

Solution:

This problem can be solved as given below:

Find the time required to design 1 painting.

Time required to design 6 paintings
= 12 days

Time required to design 1 painting
= 12 ÷ 6 = 2 days

So Mr. Parker will take 2 days to design 1 painting.

Example 2:

5 mice can eat a slice of cheese in 8 days. If 3 more mice join, how long will they take to finish the slice of cheese together?

Solution:

This problem can be solved using the following steps:

Step 1: Find the time 1 mouse takes to eat a slice of cheese.

Time taken by 5 mice to eat a slice of cheese = 8 days

Time taken by 1 mouse to eat the slice of cheese = 8 × 5
= 40 days

Step 2: Find the time 8 mice take to eat the slice of cheese.
There will be total of 8 mice if 3 more mice join them.

Time taken by 1 mouse to eat the slice of cheese = 40 days

Time taken by 8 mice to eat the slice of cheese = 40 ÷ 8
= 5 days

So they will take 5 days to finish the slice of cheese together.

Write the answer.

1. 5 workers can weed a field in 2 days. How many days will 1 worker take to weed the same field?

Answer: _____ _____
unit

2. Nikhil can fill 8 buckets in 24 minutes. How many minutes will he take to fill 1 bucket?

Answer: _____ _____
unit

Write the answer.

3. 4 robots can assemble 3 motorbikes in 6 hours. How many motorbikes can these 4 robots assemble in 14 hours?

Answer: ____ _____
unit

4. 4 people can use a water tank in 2 days. How many days will 1 person take to use the same tank?

Answer: ____ _____
unit

5. Mr. Smith can sew 15 dresses in 30 days. How many days will he take to sew 1 dress?

Answer: ____ _____
unit

6. 3 workers can dig a hole in 8 hours. If 1 more worker joins, how long will they take to dig the hole together?

Answer: ____ _____
unit

7. 5 students can do 4 assignments in 8 hours. How many assignments can these 5 students do in 12 hours?

Answer: ____ _____
unit

8. 3 tractors can weed a field in 6 hours. How many hours will 1 tractor take to weed the same field?

Answer: ____ _____
unit

9. Stephen can write 10 pages in 80 minutes. How many minutes will he take to write 1 page?

Answer: ____ _____
unit

10. 3 friends can decorate a hall in 7 hours. If 4 more friends join, how long will they take to decorate the hall together?

Answer: ____ _____
unit

11. 7 engineers can make 2 projects in 14 days. How many projects can these 7 engineers make in 35 days?

Answer: ____ _____
unit

12. 4 people can mow a field in 2 days. How many days will 1 person take to mow the same field?

Answer: ____ _____
unit

8.3 Concepts of Work and Time—3 (*)

Example 1:

A microwave oven takes 35 minutes to bake a cake. How long will it take to bake 4 cakes?

Solution:

Time to bake 1 cake = 35 minutes
Time to bake 4 cakes = 35 × 4
                                = 140 minutes

So it will take 140 minutes to bake 4 cakes.

Example 2:

4 carpenters take 12 hours to make a bed. How many hours will 1 carpenter take to make the same bed?

Solution:

Time taken by 4 carpenters to make a bed
                            = 12 hours
Time taken by 1 carpenter to make the
            same bed = 12 × 4
                            = 48 hours

So 1 carpenter will take 48 hours to make the same bed.

Example 3:

A group of 24 people can build a 200 ft. wall in 1 week. How many people do we need to build 500 ft. wall in the same time?

Solution:

This problem can be solved using the following steps:

Step 1: Find the number of people required to build a 1ft. wall in 1 week.

Number of people required to build a 200 ft. wall = 24

Number of people required to build a 1 ft. wall = 24 ÷ 200

$$= \frac{24}{200}$$

Step 2: Find the number of people required to build a 500 ft. wall in the same time.

Number of people required to build a 1 ft.

$$\text{wall} = \frac{24}{200}$$

Number of people required to build 500 ft.

$$\text{wall} = \frac{24}{200} \times 500$$

$$= \frac{\overset{12}{24} \times 5}{\underset{1}{2}}$$

$$= 12 \times 5 = 60 \text{ people}$$

So we need 60 people to build a 500 ft. wall in the same time.

Write the answer.

1. Charles takes 6 hours to make a special piece of wall art. How long will he take to make one-third of the wall art?

Answer: ____ _____
unit

2. A group of 6 ants can collect 180 grains in 1 day. How many ants do we need to collect 300 grains in the same time?

Answer: ____ _____
unit

3. 5 workers take 8 hours to make 150 bricks. How many hours will 1 worker take to make those bricks?

Answer: ____ _____
unit

4. A camera takes 3 seconds to capture a photo. How long will it take to capture 14 photos?

Answer: ____ _____
unit

5. Kelvin takes 3 hours to complete a special task. How long will he take to complete two-thirds of the task?

Answer: ____ _____
unit

6. A group of 6 women can water 100 plants in 1 hour. How many women do we need to water 200 plants in the same time?

Answer: ____ _____
unit

7. Daniel takes 8 days to write a magazine. How long will he take to write three-fourths of the magazine?

Answer: ____ _____
unit

8. A vacuum cleaner takes 2 hours to clean a floor. How long will it take to clean 3 floors?

Answer: ____ _____
unit

9. 5 workers take 15 days to repair a road. How many days will 1 worker take to repair the same road?

Answer: ____ _____
unit

10. A group of 20 worker bees can collect 200 spoons of honey in one week. How many worker bees do we need to collect 400 spoons of honey in the same time?

Answer: ____ _____
unit

8.4 Review of Chapter 8 (*)

Write the answer.

1. A tap takes 4 hours to empty 1 tank. How long will it take to empty 2 tanks?

    Answer: _____ _____
    unit

2. 6 rabbits can drink a bucket of water in 5 days. If 4 more rabbits join, how long will they take to drink the bucket of water together?

    Answer: _____ _____
    unit

3. 3 tailors take 8 hours to make some clothes. How many hours will 1 tailor take to make those clothes?

    Answer: _____ _____
    unit

4. 2 people can clean a room in 20 minutes. How many people will clean the room in 8 minutes?

    Answer: _____ _____
    unit

5. 3 students can prepare a seminar report in 4 days. How many students will prepare the report in 2 days?

    Answer: _____ _____
    unit

6. A group of 4 girls can make 60 dolls in 1 day. How many girls do we need to make 180 dolls in the same time?

    Answer: _____ _____
    unit

7. 7 people can run 50 miles in 1 day. How many miles can 14 people run in 1 day?

    Answer: _____ _____
    unit

8. 3 workers can weed a garden in 3 days. How many days will 1 worker take to weed the same garden?

    Answer: _____ _____
    unit

Write the answer.

9. It takes 8 workers to make 1 building in a month. How many workers are needed to make 2 buildings in a month?

Answer: ____ _____
unit

10. 3 friends can fill a tank by using buckets in 5 hours. If 2 more friends join, how long will they take to fill the tank together?

Answer: ____ _____
unit

11. A group of 12 robots can assemble 120 bikes in 1 week. How many robots do we need to assemble 200 bikes in the same time?

Answer: ____ _____
unit

12. Rob can write 2 stories in 40 minutes. How many minutes will he take to write 1 story?

Answer: ____ _____
unit

13. It takes 2.5 hours to watch a movie. How long will it take to watch 4 movies?

Answer: ____ _____
unit

14. A tiger takes 2 minutes to run 1 mile. How long will it take to run 14 miles?

Answer: ____ _____
unit

15. Mr. Woakes can eat 4 pounds of rice in 8 days. How many days will he take to eat 1 pound of rice?

Answer: ____ _____
unit

16. Charles takes 4 hours to do an assignment. How long will he take to do half of the assignment?

Answer: ____ _____
unit

9. Mixture Problems

9.1 Mixture Problems with Objects (**)

Example 1:

Two jars (A and B) have 60 candies each. One-half of the candies in Jar A are orange flavored, and two-thirds of candies in Jar B are orange flavored. How many candies in both Jar A and Jar B are orange flavored?

Solution:

You can find the answer as shown below:

Step 1: Find the total number of orange candies in Jar A.

Number of candies in Jar A = 60
Number of orange candies in Jar A
= half the number of candies in Jar A
$= \frac{1}{2} \times 60 = \frac{60}{2} = 30$ orange candies

Step 2: Find the total number of orange candies in Jar B

Number of candies in Jar B = 60
Number of orange candies in Jar B
= two-thirds of the number of candies in Jar B
$= \frac{2}{3} \times 60 = \frac{120}{3} = 40$ orange candies

Step 2: Find the total number of orange candies in Jar A and Jar B.

Total number of orange candies
= orange candies in Jar A
+ orange candies in Jar B
= 30 + 40 = 70 orange candies

So 70 candies in both the jars are orange flavored.

Example 2:

A box has 25 green marbles and 15 white marbles. What fraction of the total marbles is white? Write your answer in simplest form.

Solution:

The following information is given:
Number of green marbles = 25
Number of white marbles = 15

You can find the answer as shown below:

• Find the total marbles

Total marbles
= number of green marbles
+ number of white marbles
= 25 + 15 = 40 marbles

• Write the number of white marbles as a fraction

Fraction of white marbles

$= \frac{\text{number of white marbles} \times 100}{\text{total marbles}}$

$$= \frac{15 \times \overset{5}{\cancel{100}}}{\underset{2}{\cancel{40}}} \qquad \leftarrow \text{cancel by 20}$$

$= \frac{15 \times 5}{2} = \frac{75}{2}$

So $\frac{75}{2}$ of the total marbles are white.

Write the answer.

1. Jessica has 45 flowers. Two-thirds of them are roses. How many roses are there?

   Answer: _____ _____
   unit

2. Basket A has 20 apples and 16 oranges. Basket B has 10 apples and 14 oranges. If we mix the apples from both the baskets together, what fraction of the total fruits is apples?

   Answer: _____

3. A bag has 12 white shirts and 10 black shirts. What fraction of total shirts is white? Write your answer in simplest form.

   Answer: _____

4. George has 60 balloons. Half of them are red. How many red balloons are there?

   Answer: _____ _____
   unit

5. 2 boxes (1 and 2) have 10 bangles each. One-half of the bangles in Box 1 are red, and two-thirds of the bangles in Box 2 are red. How many bangles in both Box 1 and Box 2 are red?

   Answer: _____ _____
   unit

6. A packet has 12 black pens and 8 blue pens. What fraction of the total pens is blue?

   Answer: _____

7. 2 jars (A and B) have 52 marbles each. Three-fourths of the marbles in Jar A are green, and one-half of the marbles in Jar B are green. How many marbles in both Jar A and Jar B are green?

   Answer: _____ _____
   unit

8. Bag A has 10 red roses and 10 white roses. Bag B has 12 red roses and 18 white roses. If we mix the roses from both the bags together, what fraction of the total roses is red? Write your answer in simplest form.

   Answer: _____

9.2 Mixture Problems with Solutions (**)

Example 1:

A 60-milliliter bottle of syrup is one-third water. What is the quantity of water in the bottle?

Solution:

You can find the answer as shown below.

Quantity of syrup = 60 milliliters

Quantity of water

= one-third of (quantity of syrup)

= one-third of 60

$= \dfrac{1}{3}$ of 60

$= \dfrac{1}{3} \times 60$

$= \dfrac{60}{3}$

= 20 milliliters of water

So the quantity of water in the bottle is 20 milliliters.

Example 2:

Bottle X has 20 ml of alcohol and 80 ml of water. Bottle Y has 10 ml of alcohol and 90 ml of water. If we mix the contents of both the bottles, what fraction of the mixture is alcohol? Write your answer in simplest form.

Solution:

The following information is given:

Quantity of alcohol in Bottle X = 20 ml

Quantity of water in Bottle X = 80 ml

Quantity of alcohol in Bottle Y = 10 ml

Quantity of water in Bottle Y = 90 ml

- Total quantity of liquid in Bottle X
      = 20 ml + 80 ml = 100 ml

- Total quantity of liquid in Bottle Y
      = 10 ml + 90 ml = 100 ml

- Total quantity of liquid in both bottles
      = 100 ml + 100 ml = 200 ml

- Total quantity of alcohol in both bottles
      = 20 ml + 10 ml = 30 ml

Fraction of alcohol in the mixture

$= \dfrac{\text{total quantity of alcohol} \times 100}{\text{total quantity of liquid}}$

$= \dfrac{30 \times \overset{1}{\cancel{100}}}{\underset{2}{\cancel{200}}}$

$= \dfrac{30}{2}$ = 15 ml

So if we mix the contents of both the bottles, 15 ml of the mixture is alcohol.

Write the answer.

1. A cough syrup has 4 ml alcohol and 96 ml other liquid. What fraction of the total liquid is alcohol? Write your answer in simplest form.

Answer: _____

2. Two buckets (Buckets A and B) have 20 liters of solution in each of them. Bucket A is one-fourth liquid detergent and Bucket B is one-fifth liquid detergent. What is the quantity of liquid detergent in both the buckets?

Answer: ____ _____
unit

3. Bottle A has 20 ml of alcohol and 30 ml of shampoo. Bottle B has 15 ml of alcohol and 35 ml of shampoo. If we mix the contents of both the bottles, what fraction of the mixture is alcohol? Write your answer in simplest form.

Answer: _____

4. A 75-milliliter bottle of shampoo has three-fifths water. What is the quantity of water in the bottle?

Answer: ____ _____
unit

5. Jar A has 200 ml of kerosene and 800 ml of oil. Jar B has 400 ml of kerosene and 600 ml of oil. If we mix the contents of both the jars, what fraction of the mixture is kerosene? Write your answer in simplest form.

Answer: _____

6. A 100-milliliter bottle of honey has one-fourth water. What is the quantity of water in the bottle?

Answer: ____ _____
unit

7. Two cans (Cans A and B) have 30 liters of solution in each of them. Can A is two-thirds milk and Can B is half milk. What is the quantity of milk in both the cans?

Answer: ____ _____
unit

8. A juice packet has 50 ml alcohol and 450 ml juice. What fraction of the total liquid is alcohol? Write your answer in simplest form.

Answer: _____

9.3 Review of Mixture Problem (**)

Write the answer.

1. Two jars (Jars A and B) have 40 liters of solution in each of them. Jar A is half milk and Jar B is one-fourth milk. What is the quantity of milk in both the jars?

   Answer: ____ _____
   unit

2. Andy has 35 candies. Two-fifths of them are strawberry candies. How many strawberry candies are there?

   Answer: ____ _____
   unit

3. Box A has 14 shirts and 10 pants. Box B has 10 shirts and 6 pants. If we mix the pants from both the boxes together, what fraction of the total items is pants?

   Answer: _____

4. A bag has 10 bifold wallets and 6 belt wallets. What fraction of total wallets is bifold wallets? Write your answer in simplest form.

   Answer: _____

5. A 75-milliliter jar of cough syrup is one-third alcohol. What is the quantity of alcohol in the cough syrup?

   Answer: ____ _____
   unit

6. Jar A has 100 ml of water and 400 ml of kerosene. Jar B has 150 ml of water and 350 ml of kerosene. If we mix the contents of both the jars, what fraction of the mixture is water? Write your answer in simplest form.

   Answer: _____

7. Bill has 50 cups. Half of them are brown. How many brown cups are there?

   Answer: ____ _____
   unit

8. A juice packet has 80 ml water and 420 ml juice. What fraction of the total liquid is water? Write your answer in simplest form.

   Answer: _____

Write the answer.

9. Two aquariums (1 and 2) have 18 fish each. One-half of the fish in Aquarium 1 are goldfish and two-thirds of the fish in Aquarium 2 are goldfish. How many fish in both aquariums are goldfish?

Answer: _____ _____
unit

10. A syrup has 5 ml alcohol and 45 ml other liquid. What fraction of the total liquid is alcohol? Write your answer in simplest form.

Answer: _____

11. Two jars (1 and 2) have 55 marbles each. One-fifth of the marbles in Jar 1 are red and three-fifths of the marbles in Jar 2 are red. How many marbles in both Jar 1 and Jar 2 are red?

Answer: _____ _____
unit

12. A 500-milliliter bottle of wash liquid is one-fifth water. What is the quantity of water in the bottle?

Answer: _____ _____
unit

13. Two buckets (Buckets A and B) have 16 liters of solution in each of them. Bucket A is one-fourth phenyl and Bucket B is one-half phenyl. What is the quantity of phenyl in both the buckets?

Answer: _____ _____
unit

14. A packet has 20 blue balloons and 20 white balloons. What fraction of total balloons is blue?

Answer: _____

15. Box A has 10 red pens and 15 blue pens. Box B has 20 red pens and 10 blue pens. If we mix pens from both the boxes together, what fraction of the total pens is red? Write your answer in simplest form.

Answer: _____

16. Jar A has 20 ml of water and 80 ml of honey. Jar B has 30 ml of water and 70 ml of honey. If we mix the contents of both the jars, what fraction of the mixture is water? Write your answer in simplest form.

Answer: _____

9.4 Filling or Emptying by One Pipe (*)

Example 1:

A water pipe can fill a container in 12 minutes. How long will the pipe take to fill two-thirds of the container?

Solution:

You can consider the whole container to be 1 and find the answer as shown below.

Time the water pipe takes to fill the whole (1) container = 12 minutes

Time the water pipe takes to

fill $\frac{2}{3}$ of the container = $\overset{4}{\cancel{12}} \times \frac{2}{\cancel{3}}$

$= 4 \times 2 = 8$ minutes

It will take 8 minutes to fill two-thirds of the container.

Example 2:

A water tap can empty a tank in 40 minutes. If we start emptying a full tank, what fraction of the tank will still be filled after 25 minutes?

Solution:

You can consider the whole tank to be 1 and find the answer as shown below.

Time the water tap takes to empty the whole (1) tank = 40 minutes

Step 1: Find the fraction of the tank the tap can empty in 1 minute.

Fraction of tank empty in 40 minutes = 1
Fraction of tank empty in 1 minute = $1 \div 40$

$= \frac{1}{40}$

Step 2: Find the fraction of the tank the tap can empty in 25 minutes.

Fraction of tank empty in 1 minute = $\frac{1}{40}$

Fraction of tank empty in 25 minutes

$= \frac{1}{\underset{8}{\cancel{40}}} \times \overset{5}{\cancel{25}}$

$= \frac{5}{8}$ of tank

Step 3: Find the fraction of tank still filled after 25 minutes.

Fraction of tank filled after 25 minutes

= 1 − fraction of tank empty in 25 minute

$= 1 - \frac{5}{8} = \frac{8-5}{8} = \frac{3}{8}$ of tank

So if we start emptying a full tank, $\frac{3}{8}$ of the tank will still be filled after 25 minutes.

Write the answer.

1. A water tap can fill a drum in 20 minutes. What fraction of the drum can it fill in 10 minutes?

   Answer: _____

2. A water tap can empty a tank in 35 minutes. If we start emptying a full tank, what fraction of the tank will still be full after 20 minutes?

   Answer: _____

3. A pipe can empty a tank in 3 hours. How long will the pipe take to empty two-thirds of the tank?

   Answer: ____ _____
   unit

4. A pump can fill a tank in 40 minutes. What fraction of the tank can it fill in 15 minutes?

   Answer: _____

5. A water pipe can fill a small pool in 6 hours. How long will the pipe take to fill one-half of the pool?

   Answer: ____ _____
   unit

6. A pipe can empty a drum in 35 minutes. How long will the pipe take to empty two-fifths of the drum?

   Answer: ____ _____
   unit

7. A water pipe can fill a container in 18 minutes. How long will the pipe take to fill two-thirds of the container?

   Answer: ____ _____
   unit

8. A water tap can fill a tank in 3 hours. If we start filling a tank, what fraction of the tank will still be left to fill after 2 hours?

   Answer: _____

9.5 Filling or Emptying by Two Pipes (**)

Example 1:

Pipe A can fill 320 liters of water in 1 hour, and Pipe B can fill 450 liters of water in 1 hour. If both the pipes are opened together, how much water will be filled in 3 hours?

Solution:

The following information is given:

Amount of water Pipe A can fill in 1 hour = 320 liters

Amount of water Pipe B can fill in 1 hour = 450 liters

• Find the amount of water that can be filled by both pipes in 1 hour.

Amount of water filled by both pipes in 1 hour = 320 + 450
= 770 liters

• Find the amount of water that can be filled by both pipes in 3 hours.

Amount of water filled by both pipes in 1 hour = 770 liters

Amount of water filled by both pipes in 3 hours = 770 × 3
= 2,310 liters

If both the pipes are opened together, 2,310 liters of water will be filled in 3 hours.

Example 2:

Tap A can fill 6 liters of water in 1 minute, and Tap B can fill 8 liters of water in 1 minute. If both the taps open together, how long will they take to fill a 70-liter tank?

Solution:

The following information is given:

Amount of water Tap A can fill in 1 minute
= 6 liters

Amount of water Tap B can fill in 1 minute
= 8 liters

We can find the answer by using the following steps:

Step 1: Find the amount of water that both taps can fill in 1 minute.

Amount of water both taps can fill in 1 minute = 6 + 8 = 14 liters

Step 2: Find the time taken by both taps to fill 1 liter of water.

Time taken by both taps to fill 14 liters of water = 1 minute

Time taken by both taps to fill 1 liter of water = $1 \div 14 = \frac{1}{14}$ minutes

Step 3: Find the time taken by both taps to fill 70 liters of water.

Time taken by both taps to fill 1 liter of water = $\frac{1}{14}$ minute

Time taken by both taps to fill 70 liters of water $= \frac{1}{\cancel{14}} \times \cancel{70}^{5} = 5$ minutes

If both the taps opened together, they will take 5 minutes to fill a 70-liter tank.

Write the answer.

1. Tap A can empty 20 liters of water in 1 minute, and Tap B can empty 15 liters of water in 1 minute. If both the taps are opened together, how much water will be emptied in 5 minutes?

   Answer: ____ _____
   unit

2. Tap A can fill 300 liters of water into a container in 1 minute, and Tap B can empty 250 liters of water from the container in 1 minute. If both the taps are opened together, how much water will be filled in 1 minute?

   Answer: ____ _____
   unit

3. Tap 1 can fill 40 liters of water into a container in 1 hour, and Tap 2 can empty 50 liters of water from the container in 1 hour. If both the taps are opened together when a 60-liter tank is full, how many hours will it take to empty the whole tank?

   Answer: ____ _____
   unit

4. Pipe A can fill 10 buckets of water in 1 hour, and Pipe B can fill 12 buckets of water in 1 hour. If both the pipes open together, how long will they take to fill 66 buckets of water?

   Answer: ____ _____
   unit

5. Tap 1 can pump 50 liters of water into a container in 1 hour, and Tap 2 can empty 30 liters of water from the container in 1 hour. If both the taps are opened together when a 100-liter tank is full, how many hours will it take to empty the whole tank?

   Answer: ____ _____
   unit

6. Pump A can fill 500 liters of water in 1 hour, and Pump B can fill 1,000 liters of water in 1 hour. If both the pumps are opened together, how much water will be filled in 2 hours?

   Answer: ____ _____
   unit

Quiz

1. What is the operation keyword(s) in the following sentence?

   Edwin gained 25% of his monthly salary.

   Answer: _____

2. What operation will you use for the keywords *fraction of*?

   (a) Addition
   (b) Subtraction
   (c) Multiplication
   (d) Division

   Answer: _____

3. 1 bottle costs $2.00. Which math sentence will you use to find the cost of 10 bottles?

   (a) 10 × 2
   (b) 10 + 2
   (c) 10 ÷ 2
   (d) All of the above

   Answer: _____

4. Sonia spent $30.00 in total. She spent $10.00 on a keyboard, $5.25 on a mouse, and the rest on a bag. How much money did she spend on the bag?

   Answer: _____

5. Brian and his brother spent $92.00 in a mall. They bought 3 T-shirts that costs the same amount and bought 2 jackets for $48.00. What was the cost of each T-shirt?

   Answer: _____

6. 4 kids can clean a room in 30 minutes. 1 kid was not feeling well and could not help the others. How long did it take the other kids to clean the room?

   Answer: _____

7. Kristin runs 18 miles in 2 hours. How far will she run in 4 hours?

   Answer: _____ _____ unit

8. I am a number with 3 decimal places. My right-most digit is 1, and every digit is 2 times the digit to its right. What number am I?

   Answer: _____ _____ unit

9. When rounding to the nearest thousand, what is the smallest number that rounds to 5,000?

   Answer: _____

10. Mike's current age is three-fourths of his sister's age. His sister is 20 years old. What is Mike's age?

Answer: ____ _____
    unit

11. Mia's age is two-fifths of her father's age. Her father's age is 50. Find the difference between their ages.

Answer: ____ _____
    unit

12. Craig returned from a 7-day cultural event that ended on May 15. On which date did the event start?

Answer: ____ _____
    unit

13. Ellen's neighborhood park is 150 meters from her home. If she took 3 minutes to walk from her home to the park, what was her speed in meters per minute?

Answer: ____ _____
    unit

14. Victor ordered 1 large cake that cost $25.00. He also ordered 5 burgers. The price of each burger was $2.50. If Victor gave $40.00 to the cashier, how much money did the cashier return?

Answer: _____

15. 4 workers can paint 8 rooms in 2 days. How many rooms can 7 workers paint in 2 days?

Answer: ____ _____
    unit

16. 3 men can mow a lawn in 8 hours. If 1 more man joins, how long will they take to mow the lawn together?

Answer: ____ _____
    unit

17. 2 bags (1 and 2) have 12 shirts each. One-half of the shirts in Bag 1 are red and one-third of the shirts in Bab 2 are red. How many shirts in both Bag 1 and Bag 2 are red?

Answer: ____ _____
    unit

18. Can A has 800 ml of milk and 200 ml of water. Can B has 600 ml of milk and 400 ml of water. If we mix the contents of both the cans, what fraction of the mixture is water? Write your answer in simplest form.

Answer: _____

19. Tap A can fill a 250-liter water tank in 1 hour, and Tap B can fill a 500-liter water tank in 1 hour. If both the taps are opened together, how much water will be filled in 2 hours?

Answer: ____ _____
    unit